Molecular Motors in Bionanotechnology

Molecular Motors in Bionanotechnology

James Youell
Keith Firman

Pan Stanford Publishing

Published by

Pan Stanford Publishing Pte. Ltd.
Penthouse Level, Suntec Tower 3
8 Temasek Boulevard
Singapore 038988

Email: editorial@panstanford.com
Web: www.panstanford.com

British Library Cataloguing-in-Publication Data
A catalogue record for this book is available from the British Library.

Molecular Motors in Bionanotechnology

ISBN 978-981-4267-02-1 (Hardcover)
ISBN 978-981-4303-09-5 (eBook)

Printed and bound by CPI Group (UK) Ltd, Croydon, CR0 4YY

Contents

Preface

It has been recognised for a considerable time that nanotechnology and especially the manipulation of single molecules represent not only a major challenge but also a major opportunity for mankind. This is illustrated from early history, where the ability to manipulate "novel" materials has led to the inclusion of the name of the "novel" material in a description of the epoch during which the new technology was created — The Stone Age and the Bronze Age are the best examples, but the "industrial revolution" is similar. The manipulation of single molecules[1] (a possible definition of nanotechnology) presents a technical challenge, but the potential of using materials, produced at this scale, can be best understood from biology where most macro-sized materials are assembled using a "bottom-up" approach of organised self-assembly of single molecules. However, biological systems are usually dynamic, and many make use of molecular machines to manipulate the biomaterials they are produced from. Such machines provide ideal models of how to design nano-scale devices for nanotechnology and some will provide the machines for nanotechnology.

The writing of this book was, in part, inspired by our work with a specific molecular motor — the Type I Restriction-Modification enzyme EcoR124I[2] — and, in particular, our attempts to produce a nanoactuator device incorporating this motor. In addition, K.F.'s involvement in the "Productive Nanoscience Road-mapping exercise" in the USA and the "TESSY Project for Road-mapping Synthetic Biology" in Europe both contributed to ideas and background thoughts about the content of the book. The authors' background has also greatly influenced the content of the book, with J.Y. having a stronger interest in structural information and structure–function relationships, while K.F. has a more chemistry-driven interest in both function and organisation of molecular machines. Our original aim was an overview of molecular motors that would link to some novel ideas about how they might be used in bionanotechnology. However, several recent reviews[3–6] have also influenced the content of the book and led to a different style and level of content from that first imagined. This final version is much longer than first imagined, but, we hope, provides an important overview of the various molecular motors that are well understood, or reasonably well described, and detailed in a way that helps understand structure–function relationships between the motors. However, we recognise that there can be an over-emphasis on the means for classifying biological systems. We are aware that this book suffers from this problem, and too much emphasis on such classifications can be misleading. Nature does not categorise, but we felt that the clarification of the classification of some systems was overdue.

Acknowledgements

K.F. would like to thank Eric Drexler for his supportive comments about K.F.'s research, which led to the initial idea of a review and then this book. Much of the work described by the authors, from their own research, was supported by two EC-funded projects: MOL SWITCH and BIONANO-SWITCH.

Chapter 1

Mode of Activity of Molecular Motors

A simple dictionary definition of a motor is "*something, such as a machine or an engine that produces or imparts motion*".* This definition works well in the macroscale world that we normally interact with. However, at the molecular scale, there is an immediate problem with this definition — the influence of thermal or Brownian motion. Brownian motion is random and can hardly be seen as motor activity; although, as we detail below it makes an important contribution to how a molecular motor works. The influence that Brownian motion has on the way a molecular motor works is elegantly summarised by Oster and Wang[7] with the phrase "*In such an environment you would not need to even pedal your bicycle: you would simply attach a ratchet to the wheel preventing it from going backwards and shake yourself forwards!*". Therefore, we should rewrite the definition in a way that eliminates the problem of thermal motion and perhaps the simplest way is "*something, such as a machine or an engine that produces or imparts directional motion*". Before discussing the wide variety of molecular motors that have been chemically synthesised, or exist in nature, it is important to describe the ways that motor activity can occur and discuss some of the problems molecular motors must overcome.

*The American Heritage® Dictionary of the English Language, Fourth Edition copyright © 2000 by Houghton Mifflin Company. Updated in 2009. Published by Houghton Mifflin Company.

Molecular Motors in Bionanotechnology
James Youell and Keith Firman

ISBN 978-981-4267-02-1 (Hardcover), 978-981-4303-09-5 (eBook)
www.panstanford.com

Almost all chemical reactions are described by equations that detail the contribution of thermal motion towards the required energy for overcoming the energy barrier between chemical states (usually through an activated state for the chemical involved), and these equations provide a means for calculating (measuring) the rate constant. For biological macromolecules, thermal motion allows the molecules to adopt conformational changes that are separated by significant energy barriers, and these changes in conformation can often be measured in solution. Therefore, for directional motion, there needs to be a significant change to the energy landscape, separating the conformational states of the molecule, which also provides a bias towards a specific state such that the energy provided by thermal noise is just insufficient to allow the change back to the initial state, but sufficient to allow adoption of the preferred state for motion. The consequence of this is that thermal noise provides a significant contribution to motion, but fuel consumption biases the energy landscape in a way that overcomes the random nature of motion that thermal energy alone would produce.

1.1 Power-Stroke Model of Motion

One can imagine two possible models for describing the type of motion produced by molecular motors, and these two models are often used to describe individual motors.[7] The first model is the "power-stroke", which is a viscoelastic model in which the motor is in a "tense" state following fuel consumption and moves to a "relaxed" state, through a number of conformational changes that occur in small (Å) steps, in a viscoelastic manner.[8] A model of a pure "power-stroke" is a charged rod (such as negatively charged DNA) passing through a polar trans-membrane pore under the influence of an applied electrical potential. The electrical potential drives the movement through the pore over a series of very small steps based on the distribution of the charge.

1.2 Brownian Ratchet Model of Motion

The second model is a Brownian ratchet in which the motor rectifies diffusion displacements over larger (nm) distances. A pure model of such a ratchet would be a charged rod (as above) passing through a hydrophobic membrane pore in the presence of a high concentration (on the entry side of the membrane) of counter-charged ions, which bind and neutralise the charges on the rod allowing it to enter the pore. The low concentration of ions on the other side of the pore allows dissociation, but not rapid rebinding of the ions, therefore preventing backward movement of the rod. This is the process of rectification and involves attractive bonds between the solvent and the charges on the rod following passage through the pore.

Brownian motors and ratchets[9] can be described as motors that provide such a biased directionality to the random fluctuations produced by thermal motion. These motors make significant use of thermal motion during their motor activity,[8] usually through such a directionally biased conformational change in the motor protein. In effect the steps that lead to directional motion are one-dimensional thermally activated transitions. However, directed motion also requires asymmetries within the system[10,11] and a key aspect of such changes to symmetry, within molecular motors, is the presence of chiral elements.[12–14]

An example of such motion, driven by the chemical changes of ATP hydrolysis, is the translocation of DNA through a viral packaging motor into the viral capsid head, against the internal forces produced by the increasing pressure produced by this packaging. Both the change in available charge following hydrolysis and the chirality of the protein–fuel interactions provide a conformational change, which in turn influences the position of the motors on the DNA and leads to directional transport of the DNA through the motor complex. The most important aspect of this motion is the coupling of the chemical reaction of ATP hydrolysis with the conformational switches in the protein. This coupling provides a mechanism for directionality; although, as seen with ATP synthase, the chemical reaction is reversible and consequently so should be the directionality, which leads to ATP synthesis when the motor is mechanically wound in a specific direction. If ATP hydrolysis was in an equilibrium state, there would be no motion and, therefore, it is the value of ΔG for ATP hydrolysis, which must be greater than the energy required to "push" the motor backwards, that provides actual directionality.[8] This energy flow also drives directionality, and the chemical reaction can only proceed in a direction that results in chemical-free energy decrease.[15] Mathematical modelling of such processes[15–18] shows that without thermal motion, a molecular motor would be trapped at a local minimum.

A key aspect of the above models for directed molecular motion is the need for fuel to drive the system, and molecular motors can use a variety of different fuels, including chemical fuel (as mentioned above for ATP), photons, and electrochemical energy. One ATP molecule is equivalent to approximately 20–24 kT of which ~8–9 kT is enthalpic, while the rest is entropic.[7] Hydrolysis of ATP in most motors is stored as 6–10 kT of elastic energy (although this elasticity is not flexibility of the binding site as this would waste energy). An important aspect of using chemical energy is that the fuel will produce by-products that will contaminate the system unless removed (as is the case for biological molecular motors).[19]

1.3 Motor Efficiency

One can readily imagine that there are two very different forces that act on molecular motors:

(a) Stall forces, which are often experimentally employed to determine the forces generated by a motor and can be illustrated by the application of a drag using an optical tweezer system.[20] These forces are often known as conservative forces and produce very different behaviour from the motor.

(b) Viscous forces in solution can be likened to friction for macroscale motors but are actually very different stochastic forces. Because molecular motors have negligible inertia and, as described above, the motor is mostly driven by Brownian force, the instantaneous velocity changes direction very rapidly and the absolute value can be of orders of magnitude greater than the average velocity of the motor.[21]

Stall forces applied to a molecular motor usually stop the motor and the chemical reaction that drives the motor (although there are cases where processivity of the motion is altered by such forces[22,23]). For a tightly coupled motor, where the step size is independent of the load force that produces stalling, and close to the stall point, the thermodynamic efficiency of the motor can approach 100%.[24,25]

However, without an external load, where the stochastic viscous forces apply, the motor will behave very differently. The interactions that produce this force are intermolecular, which are particularly important in polar solvents such as water, and as the size of the motor approaches the nanometre scale and the geometry of these events becomes very significant at these small scales.[26] Efficiency under these circumstances is usually determined as Stokes efficiency[15] — the ratio of "mechanical performance" to energy supply and this can also approach 100% when the driving force is almost constant.

Chapter 2

Chemical Motors

As detailed in Chapter 1, molecular motors are subject to different forces and parameters when compared to motors and devices that operate at the macroscale. In particular, these molecular-scale motors have no inertia and are subject to continuous Brownian motion that is random in nature. For small motors, where their size is similar to intermolecular distance, friction is "governed" by intermolecular interactions and the forces associated with such interactions. Directional motion requires three components[27] and can be best described as diffusion process along a specific dimension of a potential energy surface[28] where motion is achieved by successive raising and lowering of these energy barriers:

1. A randomising element (usually provided by Brownian motion).
2. An energy input, which is required to obey the second law of thermodynamics.
3. An asymmetry in the energy potential along the direction of movement, which can be manipulated by the operator to provide the directional motion.

A chemical motor can be described as a molecule capable of converting energy into such directed and organized motion. The first descriptions of such motors appeared simultaneously in *Nature* in September 1999 in two papers that presented rotary chemical

Molecular Motors in Bionanotechnology
James Youell and Keith Firman

ISBN 978-981-4267-02-1 (Hardcover), 978-981-4303-09-5 (eBook)
www.panstanford.com

motors capable of unidirectional motion.[29,30] Each motor used a different energy source to provide the motion, in one case light and in the other case chemical (a review of the broad range of chemical motors was produced by Kay *et al.*[3]).

2.1 Rotary Chemical Motors

There is little doubt that the driving force behind the design of chemical molecular motors are the beautiful and exquisitely efficient motors of biological systems and, as will be described in detail later, the archetypical rotary motor of nature is ATP synthase — this system is mentioned in many of the papers describing chemical motors.

2.1.1 Helicene-Based Motors

The first of the two letters to Nature[29] describing rotary chemical motors demonstrated a motor capable of unidirectional rotation through 120° around a single chemical bond, driven by chemical energy. This motor was constructed of two sections around a chiral carbon (Fig. 2.1), one section is a 3-blade triptycene (coloured grey in Fig. 2.1), while the other is a helicene (coloured a darker shade in Fig. 2.1).

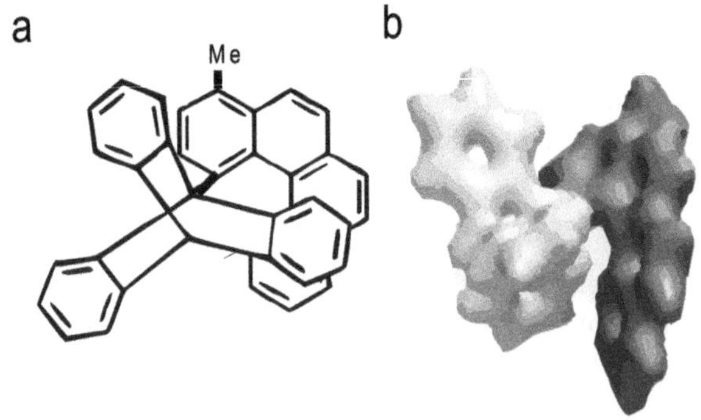

Figure 2.1 A triptycene-based rotary motor driven by chemical energy. Showing (a) the chemical structure of triptycyl[4]helicene and beside that (b) an electron-density representation of the conformation of lowest energy for the triptycyl[4]helicene molecule. Adapted from bibliography reference [29].

The unidirectional behaviour is accomplished by reversible use of a "tether" between these two components using carbonyl dichloride to create a covalent bond, which makes one direction of rotation energetically favourable. Rotation requires ~25 kcal mol^{-1}, which is very high compared to normal rotation around a carbon–carbon bond (3–5 kcal mol^{-1}). However, this limited rotation is slow (~3 h) and represents only a single step for the motor — the covalent bond must be broken and a new bond formed between the next aligned components of the two sections. As the authors themselves explain,[31] "*The next step, however, is now clear: To achieve repeated rotation by modifying (the motor) so that each blade of the triptycene is ready to be selectively armed at the appropriate time*". This motor is quite interesting compared to some other motors as it is not a Brownian Ratchet,[9] but is more like a "brake" system, where the brake prevents movement in one dimension. The use of carbonyl dichloride as the fuel for the motor, through the creation of a non-identical energy landscape for the direction of motion, can be likened to the use of high-energy fuels (e.g., ATP) by biological motors (Fig. 2.2).

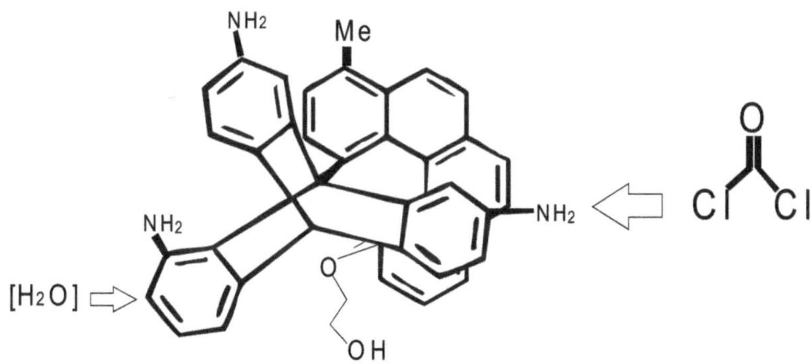

Figure 2.2 A proposal for sequential firing of rotation by carbonyl dichloride. Proposed mechanism for a continually rotating molecular motor involving repeated delivery of carbonyl dichloride ($Cl_2C{=}O$) to the amino "firing group" followed by cleavage of the urethane only after 120° rotation. Adapted from reference [31].

Kelly *et al.*[32] described the construction of a mechanism for repeated firing of such a motor in which they attached an amino group to each blade of the triptycene group and directed a phosgene molecule to the amino group through a "targeting" 4-(dimethyamino)pyridine group. However, the motor did not rotate as expected presumably due to unexpected inhibition or steric hindrance between the component groups.

2.1.2 A Chemical Motor Incorporating a Chiral Carbon–Carbon Double Bond

Perhaps one of the best described chemical, rotary motors were built around two chiral carbon atoms at either end of a carbon–carbon double bond (Fig. 2.3). This carbon–carbon double bond separates two "sets" of ring structures, which, in the motor, will rotate around the double bond.[30]

As can be seen from Fig. 2.3, the rotation is a four-step process in which the motor rotates through 360° using UV light and heat to drive the isomerisation of the molecule. The thermally induced changes are irreversible due to steric hindrance of the equatorial methyl groups, which provides a directed rotation of the motor. Continuous rotation can be achieved through an appropriate use of the wavelength of light and variation in temperature.

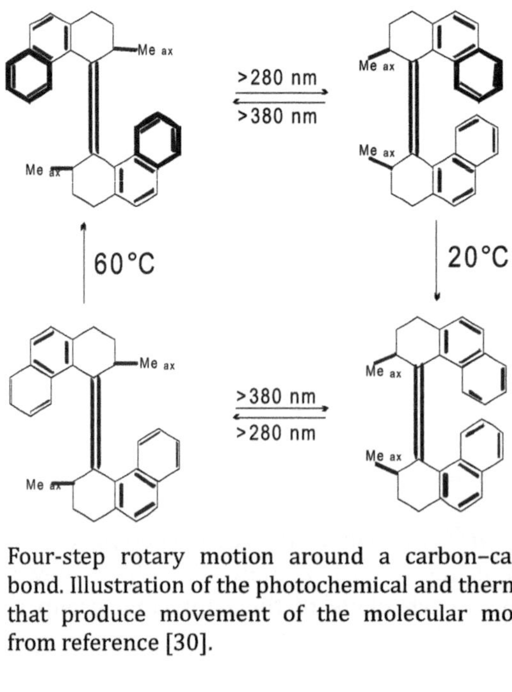

Figure 2.3 Four-step rotary motion around a carbon–carbon double bond. Illustration of the photochemical and thermal processes that produce movement of the molecular motor. Adapted from reference [30].

While the motor described above provides a first example of a genuine, working, unidirectional, rotary molecular motor, it is not easy to imagine how to obtain useful work from the motor without surface attachment; unfortunately, the symmetrical nature of the construct makes this difficult. Therefore, Koumura *et al.*[33] described the design and synthesis route for variants of the original motor that were to provide the mechanism for surface attachment as described by van Delden *et al.*[34] Attachment was achieved through two thiol-functionalised C_8 "legs", which also provided a suitable spacer between the motor and the gold surface to prevent direct, electronic interactions between the chromophore and the gold (as illustrated in Fig. 2.4). This device is a very important step towards a chemical mimic of the biological rotary motor ATP synthase.

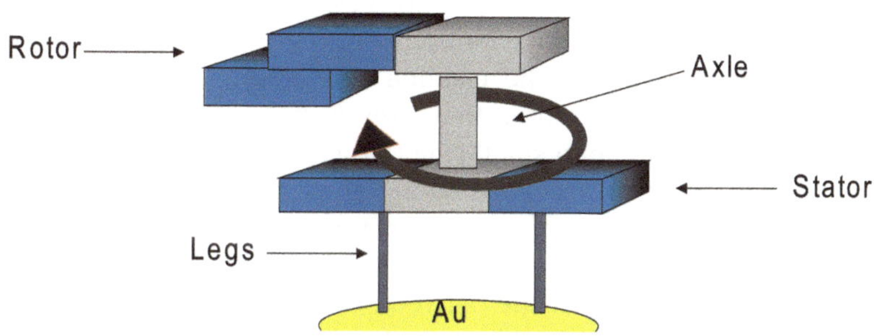

Figure 2.4 A surface-attached motor acting as a propeller. Illustration showing the design of the surface-bound molecular motor, showing the central axis around which rotation occurs and the stator connected to the gold surface via two legs consisting of a "C_8" structure bound to the cold via a sulphide linkage. Adapted from reference [34].

However, this motor is slow with a half-life for the isomerisation prior to thermal rotation being some 2,400 s. Interestingly, some of the changes required to introduce the required asymmetry detailed above, also lowered the energy barrier for rotation, which suggested the possibility of improving the speed of the motor. A variety of changes were explored by ter Wiel *et al.*,[35] including changes from six-membered ring structures to five-membered rings, which were found to dramatically lower the energy barrier to rotation and increase the speed of the motor. Perhaps of even greater significance was that these variants were much more efficiently synthesised with yields of ~76%, and they provided continual rotation at room temperature. The adaptability of this system was further explored by the addition of further groups to the stator of the motor, with *o*-xylyl groups providing steric hindrance to the rotation and providing a light-driven gearbox. Unfortunately, precise control of this system was difficult, but the principle of a rotating chemical motor has been shown and the potential for their future use in macroscale devices remains a distinct possibility.

2.1.3 Rotaxane-Based Rotary Motors

Rotaxanes are examples of physically linked molecules that are not bound to each other through covalent bonds, and they represent a group of very adaptable molecules that can be used as both rotary and linear motors.[14] In fact they are best described as Brownian Ratchets in which directional motion is achieved through successive raising and lowering of the maxima and minima on the energy landscape.[27] A rotary motor that uses a rotaxane is describe in Fig. 2.5 and consists of a small ring that can move around a larger ring, which has two specific binding sites that attract and bind the smaller ring. By manipulation of the sites, through chemical modification, the population of one site can be made energetically more favourable. Movement between the two sites is driven by random Brownian motion; however, the order of four chemical reactions, which govern the location of the smaller ring, leads to directed motion.

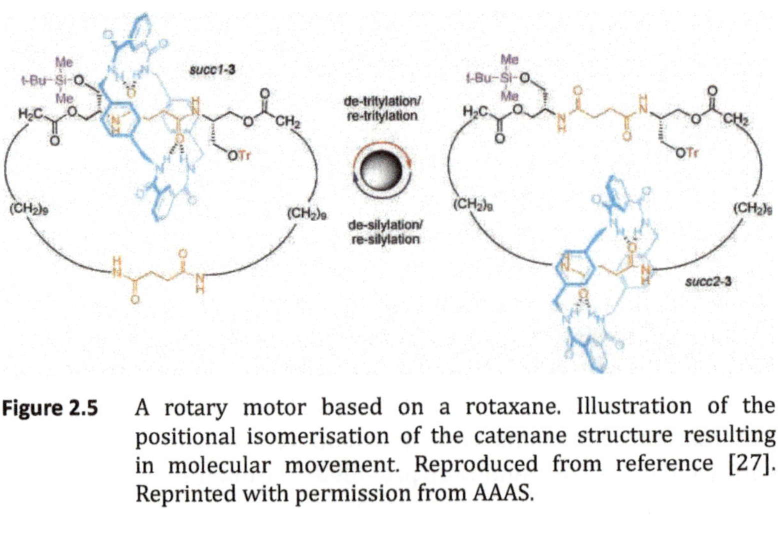

Figure 2.5 A rotary motor based on a rotaxane. Illustration of the positional isomerisation of the catenane structure resulting in molecular movement. Reproduced from reference [27]. Reprinted with permission from AAAS.

In fact, rotaxanes are capable of three possible motions[14]:

1. Axle rotation within the wheel (the smaller ring).
2. The wheel can move along the axle (a molecular shuttle).
3. The angle between the wheel and the axle can change in a pivotal fashion.

In addition, the motion of the wheel along the axle can be governed not only by chemical energy as described above, but also by light and by electrical energy. The use of an external electric field, in fact an AC current, is of particular interest as this represents an important opportunity for interfacing such a device with existing electronic systems. The AC field will produce both rotation and pivoting, and these movements seem to be governed by H-bonding interactions between the wheel and the axle.[14] Directionality of such systems also requires chirality (c.f. the presence of a chiral, helical proton channel in the biological molecular motor ATP synthase) and one well-described system uses a sulphonamide group on the wheel to introduce this chirality.

The potential of these motors in nanotechnology remains uncertain, but they represent an important step towards linear motors and the demonstration of artificial muscles.

2.2 Linear Chemical Motors

Linear motors and linear motion can be seen as an active transport system, which can easily be imagined as a means for producing useful work from a molecular motor. In addition, the muscle system that makes use of the myosin motor represents an ultimate aim for most linear motor development, and the ability of myosin to produce macroscale movement through multi-parallel activity is an important driver for work in this area.

2.2.1 Rotaxane-Based Linear Motors

Serreli *et al.*[37] described a linear chemical motor based on a rotaxane, driven by light that, like the rotary motor described above, allows movement of the smaller ring (wheel) between two locations on an axle, but the axle is linear in this case. The wheel cannot escape off the end of the axle because of the presence of two bulky "stoppers" at each end of the axle.

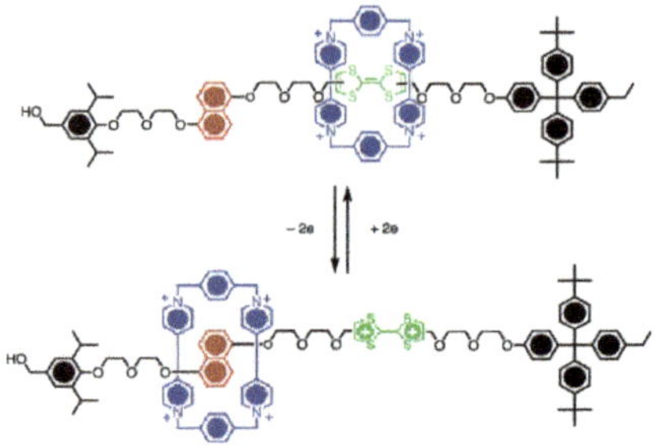

Figure 2.6 A rotaxane-based linear chemical motor. Structural representation of the bistable[2]rotaxane motor showing the two positions of the ring structure. Reprinted with permission from reference [36]. Copyright 2005 American Chemical Society.

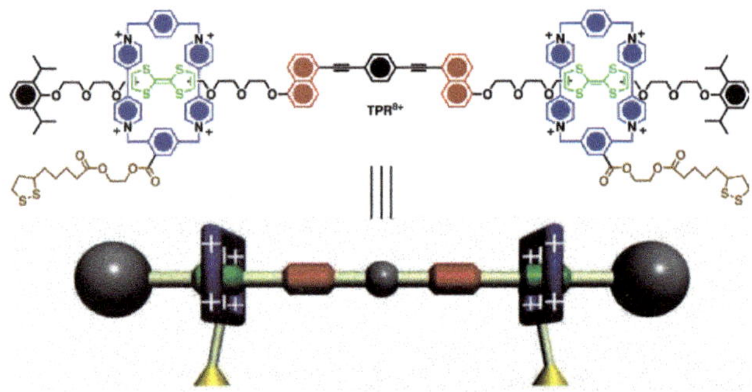

Figure 2.7 A tethered molecular shuttle. Structural and graphical representation of the disulphide-tethered molecular motor used to produce an artificial muscle. Reprinted ("adapted" or "in part") with permission from reference [36]. Copyright 2005 American Chemical Society.

The interesting aspect of this motor is the use of photosensitisers on the wheel, which signal the position of the wheel on the axle and open or close a "gate" to enable movement away from an equilibrium situation. This system can be seen as an intelligent device that signals molecular position and allows logical, directed motion.

While such devices represent a molecular shuttle (first described by Anelli *et al.*[38] as a prototype design using redox potential to drive movement — Fig. 2.6), the first step towards a potentially useful device was the description of an artificial muscle by Liu *et al.*[36] Once again this system was driven by redox potential, but, in turn, this could be derived from an applied voltage; however, the most important feature of this motor was the fact that the wheels could be tethered (Fig. 2.7), as could the axle itself. Such a situation allows the motor to exert a force against the surface and as a consequence to act as a molecular-scale artificial muscle (Fig. 2.8).

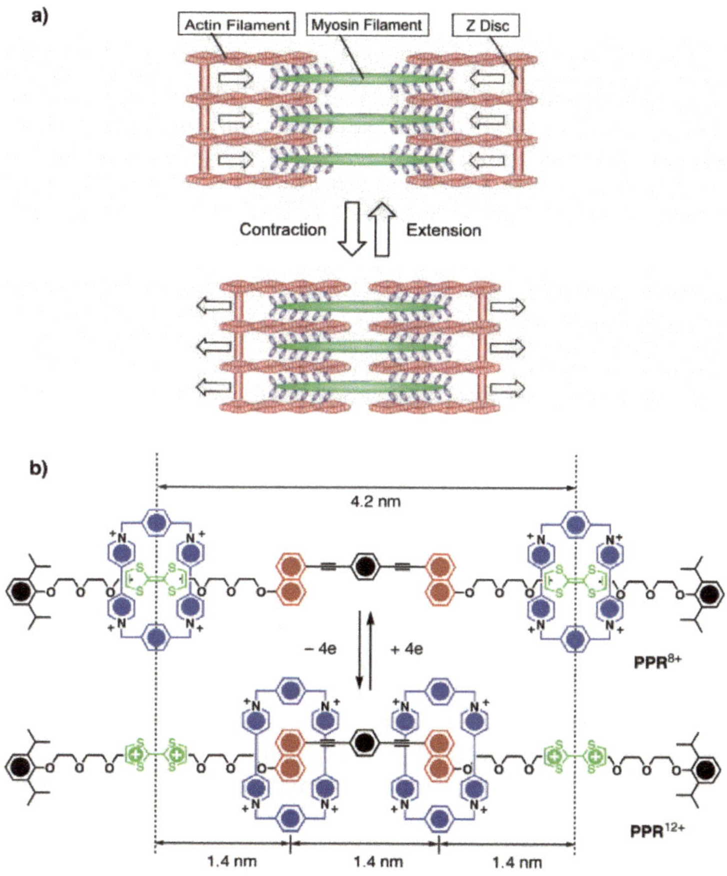

Figure 2.8 A rotaxane dimer acting as an artificial muscle. Graphical representation (a) of the contraction and extension of skeletal muscle; (b) structural formulae of the extended (top) and contracted (bottom) versions of the artificial muscle motor. Reprinted with permission from reference [36]. Copyright 2005 American Chemical Society.

This motor was assembled onto a gold-coated micro-cantilever of an AFM as a self-assembling monolayer (SAM), and the AFM was able to clearly show bending of the cantilever, by the attached motor, which is the first clear demonstration of a chemical molecular motor providing sufficient work to be classed as a NEMS (nanoelectromechanical systems) device.[39] The motion produced is in fact 5-orders of magnitude bigger than the motor itself and is a strong indicator of the future possibilities for the use of chemical molecular motors.

Other molecular systems have been described for the production of rotaxanes, but one system of particular interest is based on the use of cyclodextrins for the production of supra-molecular machines.[40] These molecules have several useful characteristics that make them interesting for the design of artificial machines and devices — they have a well-characterised ring structure suitable for rotaxane formation, and they have an ability to assemble around both small molecules and large polymeric molecules, which provides a means for their use as a molecular shuttle.

2.2.2 DNA-Based Molecular Motors

An alternative version of a chemical motor is a DNA-based biochemical motor.[41] DNA is a superb material for molecular construction and has been used to produce complex 2-D patterns using "DNA origami"[42,43] and 3-D structures that self-assemble from easily defined DNA sequences.[44–48] This ability to construct a wide range of shapes and layouts using DNA also allows the self-assembly of nanoparticles[49] on these arrays providing a rigid scaffold, which has recently led to the demonstration of single molecule chemistry on these patterned DNA structures.[50] It is possible to create stable intermediate states with DNA, which would provide the mechanism for ordered motion.[51] Therefore, it is easy to design moving devices from self-assembled DNA oligonucleotides.

One of the earliest descriptions of such a device involved a simple chemically induced transition in the DNA structure (Fig. 2.9) — a structural flip between B- and Z-form DNA — and this movement was detected by placing fluorophores on the ends of the DNA molecules and measuring fluorescence resonance energy transfer (FRET) signals.[52] This was quickly followed by a DNA-fuelled device that could act as a simple tweezer system,[53] which consisted of two rigid dsDNA molecules connected through a flexible ssDNA "hinge" and used other DNA molecules as the fuel to enable motion with a timescale of ~10 s per movement. This motion was again detected through fluorescence output.[54]

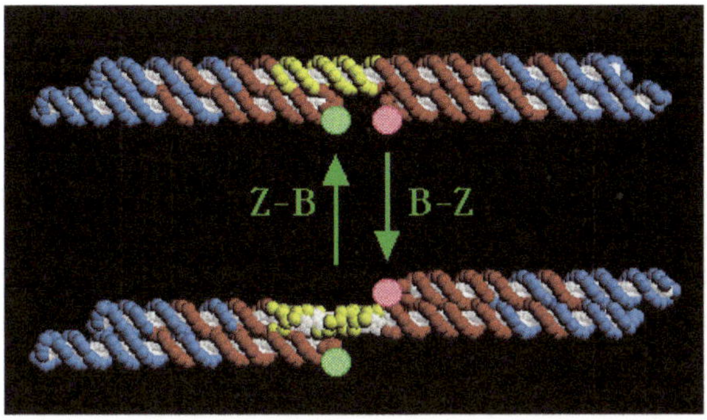

Figure 2.9 A DNA motor based on B to Z transitions. This simple chemical motor depends upon the properties of DNA,[52] which is capable of adopting different conformations dependent upon the buffer composition,[55] the motor action of the B–Z transition was monitored using FRET fluorescence from the two fluorophores at the end of each DNA molecule (pink and green circles). Reproduced with permission from Prof. Nadrian Seeman.

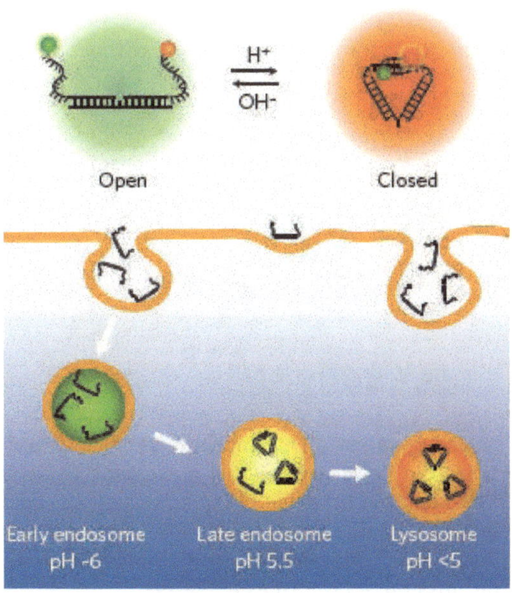

Figure 2.10 Closure of a simple DNA-based nanomachine at low pH. A pH-sensitive DNA-based nanomachine. At high pH, the structure is open, separating the two fluorophores, but at low pH, the structure closes and produces a FRET signal. This process, *in vivo*, is illustrated with endosome to lysosome development. Adapted from reference [57].

Recently, another interesting DNA-based nanodevice was described, this time sensitive to pH and is based on the formation of a proton-sensitive i-tetraplex within a DNA structure,[56] which moves the device from an open to a closed state in only a few tens of seconds (Fig. 2.10). The ends of the structure are cytosine-rich, and at low pH (5.5), these cytosines are protonated and produce two parallel-stranded duplexes that are intercalated in an antiparallel orientation. This brings together the ends of the DNA structure in such a way that the closed state resembles a triangle and allows two fluorophores at the ends of the DNA to come into close proximity, producing a FRET fluorescent signal.[57] In effect this device is a miniature probe of pH that works very effectively over the pH range 5.5–6.8, which is a physiologically important pH range.

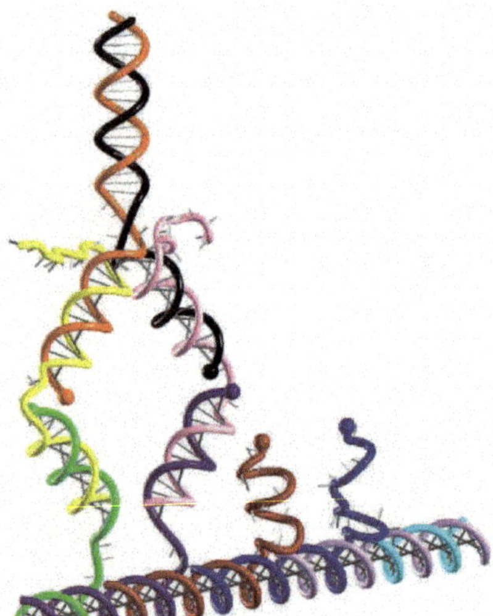

Figure 2.11 A DNA "walker". This machine is also entirely built from DNA, but in this case the movement is dependent upon hybridisation between complementary strands of DNA, and forward, unidirectional motion is ensured through the use of ligation and cleavage reactions.[59] Image reproduced with the permission of Prof. Niles A. Pierce.

However, such devices are not strictly molecular motors as their movement requires external input. While this provides a useful degree of control, truly useful motors would run constantly without external input. One such motor would involve DNA molecules that could perform walks along a track, in a manner that would be analogous to kinesin (see Chapter 4), with directionality.[58–61] Early versions of DNA walkers were still driven by addition of DNA.[62,63] The first truly useful "DNA walkers" (e.g., Fig. 2.11) made use of DNA hydrolysis either through a ribozyme-like process,[64] or through the use of a nicking enzyme[65] and run continuously along the tracks. Directionality requires destruction of the track behind the moving motor, but the steps are quite rapid (0.003 s^{-1} and 0.06 s^{-1}, respectively), limited only by the rate of hydrolysis. DNA hydrolysis provides a useful energy source and was combined with DNA ligation by Yin *et al.*[59] to produce a "DNA walker" that does not dissociate from the track — the motor is attached to the track through covalent bonds, which are cleaved by restriction enzymes and subsequently ligated using T4 DNA ligase in an ATP-dependent manner. However, as the authors themselves point out this system is a rather complicated use of two restriction enzymes and a ligase.

2.3 Potential for Use of Chemical Motors in Bionanotechnology

A variety of chemical motors, both rotary and linear, have been constructed and shown to perform "useful" work (Table 2.1). It is clear that the synthetic chemistry required is complex and needs highly specialised input and the yields of a working motor are often quite low. However, despite these problems, some of these motors have been attached to surfaces and used to create movement on the macroscale (acting as chemical versions of muscle), and it is clear such motors will be incorporated into future developments of devices for nanotechnology. However, DNA represents an important material for the construction of molecular motors as the simplicity of production of oligonucleotides (usually these are purchased) and the use of a simple set of rules that allow self-assembly of DNA devices overcome the problems associated with having to use synthetic chemistry. In addition, the fact that DNA "contains" information suggests future uses where this information is used to guide the activity of a device in a manner not unlike the ribosome. The real challenge for those developing chemical-based molecular motors is to demonstrate their use in a robust device that can operate across the micro- and macroscales, and progress towards such devices will be discussed in the final chapter of this book.

Table 2.1 Summary of available chemical motors

Type of motor	Primary function	Source of motor	Most likely use in bionanotechnology	Reference
Chiral carbon rotary motors	Light-driven rotation	B. L. Feringa Department of Organic Chemistry, Stratingh Institute, University of Groningen, Nijenborgh 4, 9747 AG Groningen, The Netherlands	A propeller or rotary generator. Conversion of light to motion in energy generation.	34
Rotaxane	Linear motion along an axis.	J. Fraser Stoddart The California NanoSystems Institute and Department of Chemistry and Biochemistry, University of California, Los Angeles, USA	Artificial muscle. A chemical lift. Electronic nanoactuator	36 66 67
DNA	Shape changes. Structural orientation of molecules	Nadrian C. Seeman Department of Chemistry, New York University, New York, NY 10003, USA	Linear motion along a DNA track. Self assembly of nanoparticles.	68 49, 69

Chapter 3

Biological Motors That Produce Rotary Motion

There are three well-described rotary biological motors:

1. The mitochondrial ATP synthase, which is responsible for conversion of a proton gradient into the "biological fuel" ATP.
2. The bacterial flagella motor, which can be either a proton-driven system or a sodium-ion-driven system.
3. The rotary DNA-loading protein found at the head of bacteriophages, which is responsible for ensuring nucleic acid, is "loaded" into the viral head.

3.1 ATP Synthase

As well as being known for its uniquely universal role at the primary source of the "*energy-currency*" of biology — adenosine triphosphate (ATP) — ATP synthase has become particularly well known for two reasons (for a recent review, see reference [70]). The first is the observation that the motor can rotate with an efficiency approaching 100%,[25] and the second is the elegant experiments that led to the visualisation of the rotary motion (Fig. 3.1).

The possibility that ATP synthase was a rotary motor was first proposed by Boyer,[71] but it was a long time before this suggestion was accepted. This eventually occurred when the atomic structure of the F_1 section of the enzyme was solved in 1994 by John Walker and colleagues.[72]

Molecular Motors in Bionanotechnology
James Youell and Keith Firman

ISBN 978-981-4267-02-1 (Hardcover), 978-981-4303-09-5 (eBook)
www.panstanford.com

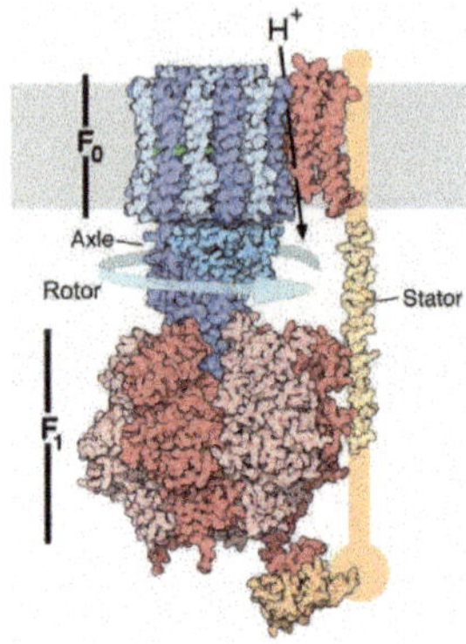

Figure 3.1 ATP synthase, a schematic model. A modified structural representation of ATP synthase from an image at the Protein Data Bank: http://www.rcsb.org/pdb/static.do?p=education_discussion/molecule_of_the_month/pdb72_1.html.

Figure 3.2 shows a schematic model of the motor and illustrates that the enzyme consists of two parts (F_0 and F_1 sections). Both parts are rotary motors; F_0 is a membrane-bound proton pump using protons (except in a few bacterial systems that use sodium ions) as the energy source for the F_1 motor[73] and the system is conserved across eubacteria, chloroplasts, and mitochondria. The eubacterial systems consist of at least 22 subunits with eight different proteins producing a complex with a molecular mass of ~350 kDa; 13 of these subunits are membrane bound and are produced from three different proteins.

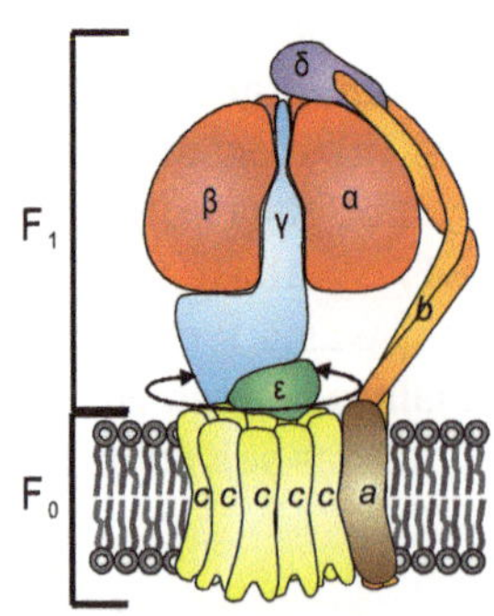

Figure 3.2 The subunits of ATP synthase. Illustration showing the various subunits that comprise the ATP synthase molecular motor: F_0 is the membrane-anchored proton pump, which drives the ATP synthase motor F_1. Adapted from the ATP synthase web site at www.atpsynthase.info.

3.1.1 Subunit Nomenclature

The subunit structure is summarised in Fig. 3.2 and, for the motor of *Escherichia coli*, the rotor consists of the γ and ε subunits of the F_1 section and the "c" subunits of the F_0 section, which form a ring within the membrane.[73] However, the nomenclature system for ATP synthase from the different systems is a little confused, with the mitochondrial system having an additional subunit, which does not occur in any other system, that is also described as the ε subunit, and the δ subunit is equivalent to the bacterial ε subunit (for a summary table see ref:[73]). The stator consists of three, alternating α and β subunits of the F_1 section and the F_0 transport-mediating "a" subunit, which is also connected to the catalytic α and β subunits by two copies of the "b" subunit (Fig. 3.2) in a stalk that is an essential component that connects the two stator elements.[74,75] In some photosynthetic bacteria and in chloroplasts, this stalk consists of two very similar proteins encoded by two genes. In the mitochondrial system, there are also an additional four subunits (e, f, g, and A6L) associated with the "a" subunit and two additional proteins in the stalk (d and F_6).

3.1.2 Mechanism of Motor Action

The F_1 section is responsible for ATP synthesis, and both F_0 and F_1 sections rotate around a common axis[76] responsible for elastic power transmission between the sections. However, the "genuine" rotary movement of the two motors is actually different. The direction of rotation depends upon the balance of the free energy released by proton movement down through F_0 versus the free energy of ATP hydrolysis. When the free energy from proton movement down into the inside of the mitochondria is greatest, F_1 is rotated against its genuine direction and ATP is synthesised. However, when the free energy released by ATP hydrolysis is greater, F_1 rotates in its genuine direction, reversing the rotation of F_0 and pumping protons out of the cell. However, to prevent a reverse reaction that might be wasteful to the cell, there is a specialised control mechanism available (in chloroplasts) based on a redox reaction that involves a disulphide bond on the γ subunit[77] and appears to act through increasing the probability of ADP inhibition of rotation.

Isolated F_1 acts only as an ATPase, hydrolysing ATP in a three-step process, which results in rotation of the motor in an anticlockwise manner (when viewed from above in Fig. 3.1). Chemical synthesis is accomplished by the β subunit, which undergoes a series of conformational changes that have been observed in the crystal structure; the subunits with bound nucleotide are "closed" structures, while the subunits that are empty are "open" to the solvent. The conformation of the β subunit depends upon the face of the γ subunit that it interacts with. Therefore, it is rotation of the γ subunit that drives the conformational changes in the β subunit and provides the energy required for ATP synthesis. The crystal structure showed three chambers in the F_1 structure carrying ATP (β-1), ADP (β-2), or no nucleotide (β-3). Recently,[78] these structures have been confirmed by an elegant set of NMR experiments in which the F_1 section was reassembled *in vitro*, with individual subunit labelling, to allow determination of a solution structure. Therefore, this suggested a mechanism of synthesis/hydrolysis of ATP involving ATP entering (β-1), movement through 120° resulting in hydrolysis of the ATP to ADP (β-2), and then another rotation of 120° releasing the nucleotide. Two amino acids, Arg^{376} and Ser^{347}, show significant positional differences between the "closed" and "open" structures of the "b" subunit and appear to provide access to the solvent and consequently the release of the nucleotide.[72] Yasuda *et al.*[79] have shown, by high-speed imaging, that the 120° step consists of roughly 90° and 30° sub-steps, each taking only a fraction of a millisecond and this time resolution has recently been further improved allowing identification of transient dwell times.[80] ATP binding drives the 90° sub-step, and the 30° sub-step is probably driven by release of a hydrolysis product. The angular position of these steps was later refined by Shimabukuro *et al.*[81] who showed that sub-steps were at 80°, which appears to be the ATP-binding step, and 40°, which appears to involve the release of ADP and Pi.[82] This release stage is influenced by binding of ATP in the adjacent β subunit (this binding has a lower affinity), and this cooperative action promotes catalysis, which is dependent upon the presence of the γ subunit.[83,84] In ATP synthesis, the 40° sub-step creates a high affinity Pi-binding site, while the 80° sub-step reduces the ATP-binding affinity producing release. The two sub-steps are separated by two reactions lasting about 1 ms, which together occupy most of the ATP hydrolysis cycle. This scheme probably applies to rotation at full speed (130 revolutions per second at saturating ATP) down to occasional stepping at nanomolar ATP concentrations. This rotational process was confirmed by fluorescence studies[85] using tryptophan mutants during steady-state ATP hydrolysis,[86] which showed that 2 moles of ADP were bound for every mole of ATP bound and that a rotation of 120° hydrolysed this ATP and that the last 120° rotation released ADP and Pi.[87] In addition, Itoh *et al.*[88] demonstrated that physical forcing of the rotation in the appropriate direction synthesised ATP (Fig. 3.3) and theoretical modelling has confirmed the experimental data described above.[89]

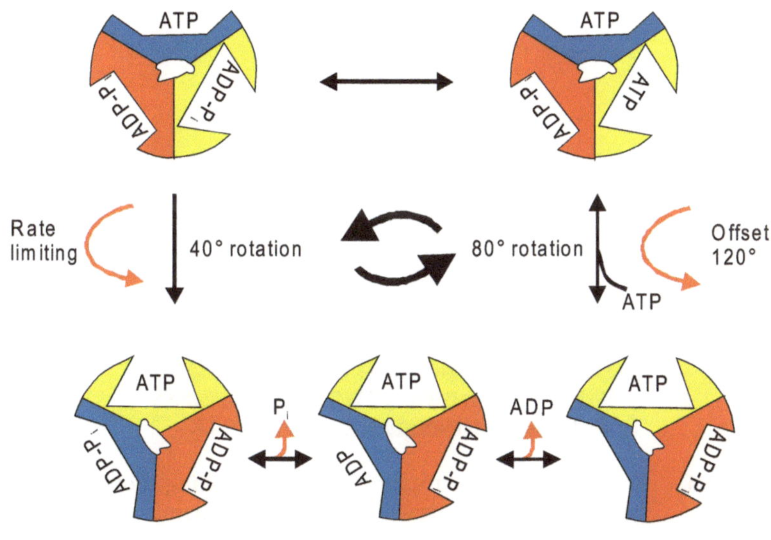

Figure 3.3 A model for ATP hydrolysis/synthesis based on rotation of the γ subunit. A model for the rotary catalysis of ATP showing (lower row) the open configuration with ADP and P_i bound and subsequently released from the motor. Adapted from reference [73].

The ATP synthase motors of bacteria and chloroplasts, but not the mitochondrial ATP synthase, are subject to autoinhibition, depending upon nucleotide conditions or proton-motive force, by one of the stalk subunits.[90,91] This inhibition is controlled by the ε subunit, which appears to adopt a highly extended structure, within the central cavity, that prevents rotation by making numerous contacts with both the stator and rotor proteins.[92]

3.1.3 Single-Molecule Analysis of Motor Activity

Direct evidence for this rotation came from observation of the rotation under a microscope. Noji *et al.*[93] immobilised the F_1-ATPase on a glass surface and attached an actin filament to the γ central axis (Fig. 3.4a) using a streptavidin-biotin linkage. The actin carried a fluorescent marker and this allowed direct observation of a unidirectional rotation of the actin filament during ATP hydrolysis (Fig. 3.4b).

Figure 3.4 Visualisation of F_1 rotation. The space-filling model of ATP synthase (a) shows attachment of an actin filament (red) to the motor through a streptavidin linkage (brown). A fluorophore is attached to the actin filament and movement of the motor, producing rotation of the actin filament, is illustrated from the captured images (b) taken at 33 ms intervals and showing an average rotation of 1.3 rev s^{-1}. Reprinted by permission from Macmillan Publishers Ltd from reference [93]. Copyright 1997.

Therefore, the F_1-ATPase has been shown to be a single-molecule molecular motor that is only 10 nm high and approximately 10 nm diameter — the smallest rotary molecular motor known.[94,95] At low ATP concentrations, where the individual steps in the rotation begin to become apparent, the torque of the motor has been measured at ~44 pN·nm and one ATP molecule is consumed per step in the rotation cycle.[95]

3.1.4 Structural Details of the Motor

NMR structural information concerning the *E. coli* F_0 motor was obtained following solvent extraction of membrane-bound proteins[96] and this section is composed of three subunits with the stoichiometry of $a_1b_2c_{9-12}$.[97] The number of "c" subunits determines the stoichiometry of $H^+ \rightarrow$ ATP synthesis,[98] and it is clear from various studies that one H^+ or Na^+ ion is transported per "c" subunit. Rastogi and Girvin[99] were able to show, from these NMR studies, that changes in pH resulted in significant conformational changes that were dependent upon protonation of a conserved carboxyl acid (glutamate in most systems, but aspartate in some bacterial systems). The "a" subunit is thought to have five trans-membrane domains that either make contact with the "c" subunit in the cytoplasmic membrane, or with a subunit in the periplasm and protons are translocated between them. The direction of rotation is linked to the passage of H^+/Na^+ through this interface, which provides a pathway through to an adjacent "c" subunit (Fig. 3.5) and results in protonation of the carboxylic acid, which, in turn, allows rotation of the "c" subunit until it encounters another "a" subunit where it de-protonates, releasing the H^+/Na^+ ion within the channel.[100] This process requires a positive charge on the stator, which forces de-protonation; this is provided by Arg^{210} that is required for coupled transport of protons[101] and this prevents "short-circuiting" of the system.

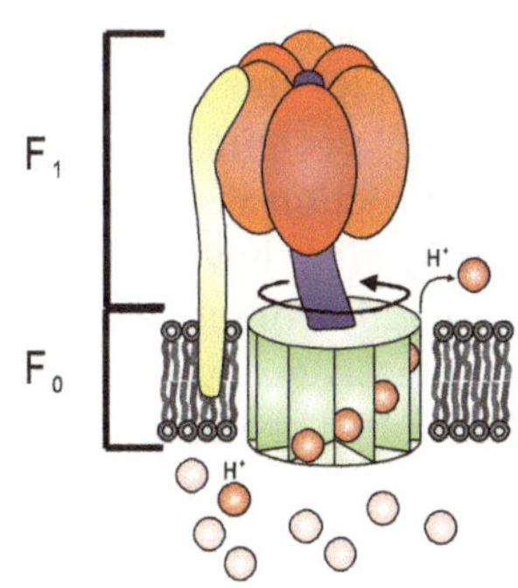

Figure 3.5 Model for proton flow and generation of movement by F_0 motor. Illustration showing the mechanism of sequential transport of a proton through the proton pump of ATP synthase, which is responsible for producing motion of the F_0 complex and the subsequent rotary motion of the F_1 complex leading to synthesis of ATP.

Several groups[102–104] were able to demonstrate rotation of the "c" ring when the γ subunit was anchored, or when the "c" ring was anchored, rotation of the γ and β subunits was observed. However, the real proof of proton-driven rotation was obtained by Fluorescent Resonance Energy Transfer (FRET) measurements, which provided positional information for the γ subunit,[105,106] and demonstrated that a proton gradient resulted in rotation in the opposite direction to that observed during ATP hydrolysis.

3.2 Bacterial Flagella Motor

Most bacteria that swim do so by using flagella, large protein filaments that extend from the bacteria (Fig. 3.6) and are usually present as six to eight individual structures each of which is termed a flagellum; although the number of flagella is tightly regulated by a negative feedback system that is linked to growth.[107,108] The flagella rotate to induce motion, which induces bundling into a stack of inter-wound flagella that rotate simultaneously, resulting in a linear, directional process.

3.2.1 Direction Switching of Flagella Motor

The direction is changed by abrupt changes to the rotation of one or a few of the flagella, from clockwise to counterclockwise rotation,[109,110] which results in unwinding of the bundle of flagella, followed by random reorientation of the bacterium and a change in direction.[111] The mechanism for this change in direction is based on phosphorylation of the CheY protein, which subsequently binds the motor and switches the direction of rotation (see later for further details). Recently, Boehm *et al.*[112] have also described an alternative system bacteria have for controlling flagella motion, which is a "brake" that can curb motor output dependent upon a second bacterial messenger c-di-GMP. The presence of c-di-GMP is signalled to the motor through the binding protein YcgR, which interacts with the proton channels in the cytoplasmic membrane to curtail motor output.

3.2.2 Structural Outline of Flagella Motor

About 40 gene products are used to construct the flagellum of *E. coli* or *Salmonella typhimurium* and these proteins are assembled from the inside outwards in a defined order.[113,114] This provides one of the best examples of self-assembly in nature and is seen by many as a model example of how future devices must self-assemble if bionanotechnology is to provide truly useful outputs.[115]

Assembly of the basal body, which consists of the MS-ring, rotor, and stator structures (Fig. 3.6) is where construction of the external flagellum initiates and the rest of the flagellum is assembled onto this basal body, through the gradually assembling tubular structures, using a Type III protein export system that is associated with the C-ring.[116]

The ring structure was named by DePamphilis and Adler[117] based on the discrete areas of the membranes with which they were associated (Fig. 3.6): M-rings for the inner membrane fraction; S-rings for supramembraneous is seen just above the inner membrane; P-rings for peptidoglycan layer; and L-rings for the lipopolysaccharide, which has affinity for the outer membrane fractions.[117–119]

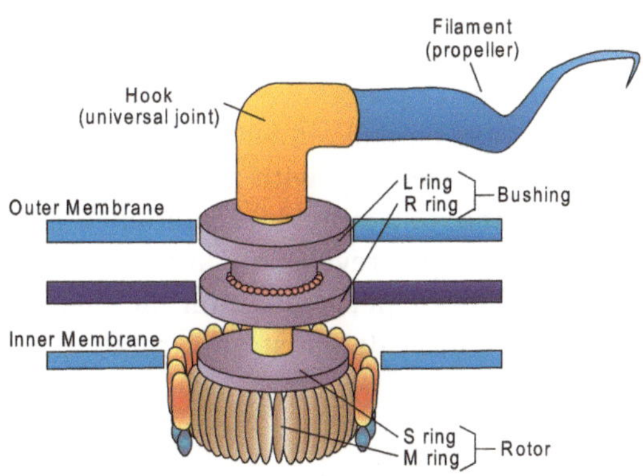

Figure 3.6 The bacterial flagellum. Illustration of the components of the bacterial flagellum embedded within the membrane of a bacterium.

Thomas *et al.*[120] working with a locked mutant of the *S. typhimurium* flagella, and Suzuki *et al.*[121] used cyro-electron microscopy (cyro-EM), single-particle imaging to visualise the rotor unit, and this work has provided some of the most detailed images of the motor. There was some dispute about the degree of rotational symmetry between the two models, but this seems to be related to the presence of FliG and the flexibility of the rotor required for movement.

In *S. typhimurium* and *E. coli*, the flagella–motor complexes have an approximate molecular mass of 10^9 Daltons and represent one of the most complex self-assembling structures known in nature. Images of the flagella assembly *in situ* are difficult with such a large complex, but cyro-EM has been applied to the smaller flagella system of *Treponema primitia*,[122] which allows better penetration of the EM beam. However, care must be employed in comparing these structures as the *Treponema* system is unusual in that the flagella itself remains buried in the periplasm and movement is produced by vibration of the cell body. Recently, electron cryotomographical observations of the flagella from all bacteria have been carried out[123] and both conserved and varied features have been observed. These observations confirmed the location of C- and M-rings and the rod. The P collar was also observed, but perhaps of most interest when comparing this structure with those obtained from isolated flagella was the presence of an associated cellular component, which appears to be the Type III protein export system.

Bacterial movement is accomplished by a spinning action of the flagella produced by the motor of the basal body driven by a proton flux; the motor is embedded between the cell wall and the cytoplasmic membrane (Fig. 3.6), with a diameter of some 45 nm, and it is composed of about 20 individual components.[124] The motor steps at about 400 times per revolution, and at low speeds, about 1000 protons are required per revolution. Studies *in vitro* with high sodium gradients have demonstrated rotation speeds up to 1,700 Hz.[125,126]

3.2.3 Subunit Nomenclature

The basal body is composed of several components that are reminiscent of the design of macroscale motors: a rotor and stator that provide the torque, a drive shaft, and a switch regulator. The protein components of the flagella and their specific localisation are illustrated in Fig. 3.7 and the key components of the motor are as follows[127–129]:

- FliF — 27 subunits comprise the MS-ring and produce the rotor.
- FliG — is located on the cytoplasmic surface of the MS-ring; however, the precise location has been difficult to determine suggesting there is a great deal of flexibility in the MS-ring,

which would fit with the movements required to generate the torque of the system.

- MotA–MotB — surrounds the MS-ring, comprises the stator, and spans the cytoplasmic membrane. MotB is also anchored to the peptidoglycan layer through its periplasmic-located C-terminal domain, which provides the proton channel for the energy flow and a stator structure against which the rotor can rotate and provide the torque. MotA binds the secondary signalling system c-di-GMP-loaded YcgR, mentioned earlier, and this reinforces electrostatic interactions between MotA and FliG, which increases drag into the system and slows the motor speed.[112]
- FliG, FliM, and FliN — are known as the "switch" proteins and are involved in the generation of torque.[130] Switching the direction of rotation is governed by a signal transduction system and involves phosphorylation of the signal protein CheY, which binds to FliM and increases the frequency of switching direction.

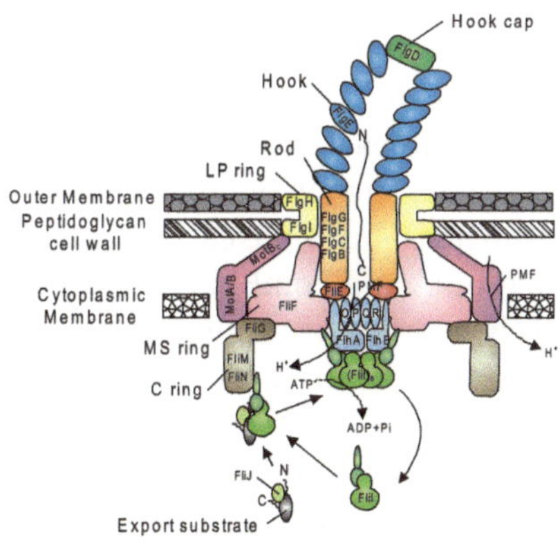

Figure 3.7 Proteins that comprise the bacterial flagellum. The hook and basal body of the bacterial flagellum showing the subunits that comprise the motor and the flagella protein export system, which is also illustrated at the base of the flagellum. Adapted from reference [115].

The stoichiometry of FliF and FliG is 1:1; the localisation of the N-terminus of FliG is close to the C-terminus of FliF; and charged residues of FliG, involved in rotation, are clustered along a prominent ridge in the structure. FliG and FliN form a "cup-like" structure known as the C-ring, which is attached to the cytoplasmic surface of the MS-ring and provides the template for assembly of the MS-ring.[130]

The main structural components outside of the cell are as follows: the 55 nm tubular hook, which provides a universal joint for the flagellum (produced from 130 copies of the *flg*E gene product) — the length of the hook is governed by FliK protein, which acts a molecular ruler[131]; in addition, there are the hook-associated proteins (*flg*K, *flg*L, and *fli*E) and the filament or propeller (20–30,000 copies of *fli*C).[132] FlgB, FlgC, FlgF, and FlgG form the rod, or drive-shaft, which connects the motor components to the hook (Fig. 3.7). However, some variations in structure and components have been observed between bacterial species.[123]

3.2.4 Mechanism of Motor Activity

The flagella motors of *E. coli* and *S. typhimurium* are driven by an inward-directed proton flux and are, therefore, not dependent upon ATP.[133] This proton flux across the MotA–MotB complex induces conformational changes that are thought to allow interactions between key charged residues of MotA with FliG.[127,134,135] However, marine bacteria and some bacteria that live at high pH use sodium ions to drive their flagella. The ion pump is believed to lie at the base of the MS-ring and again is composed of equivalents of the MotA and MotB proteins (PomA and PomB) and, as with the *E. coli* proton pump,[136] a negatively charged aspartate residue has been shown to be highly conserved and required for rotation.[137] Recent work by Terashima *et al.*[138] has investigated the requirement to maintain the conserved Asp24 in the MotB protein of *Vibrio alginolyticus*, and they were able to show that loss of Asp24 could be suppressed by introduction of a negatively charged residue in MotB (N194D). Therefore, it seems that proton flow occurs through a negatively charged pocket in the MotA–MotB (PomA–PomB) complex, and interaction with the FliG proton, through a conformational change, results in production of the motive force. There may be as many as 16 pairs of these proteins, arranged in a circular ring or rotor-like arrangement,[124] each capable of stepping one-step due to passage of a proton (c.f. F_0 ATP synthase,[139] which has also been shown to involve an aspartate residue). Interestingly, recent work has shown that it appears that the electrostatic interaction of MotA with FliG is required for the efficient assembly of the stators around the rotor,[135,140] but that proton translocation activity is not required for stator assembly.

The full structure of the FliG subunit (Fig. 3.8) from *Aquifex aeolicus*[141] has allowed direct observation of conformational changes associated with spin-direction changes, which is used to rapidly change the direction a bacterium follows, and this structure, combined with mutation studies, enabled molecular modelling techniques to describe the reversing of electrostatic charges involved in torque generation.

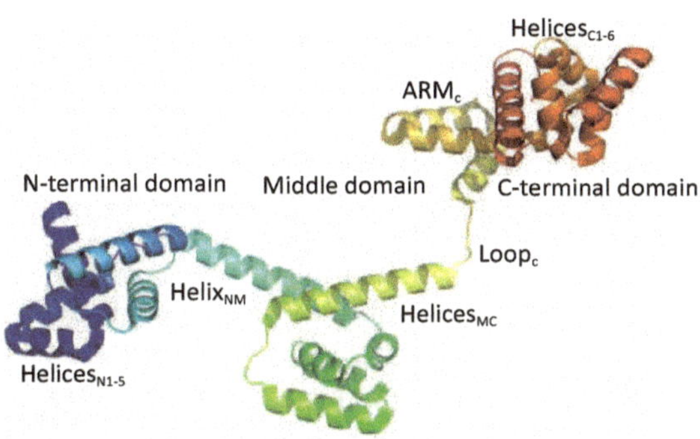

Figure 3.8 Structural overview of the FliG subunit. Graphical representation of the structure of the FliG subunit showing the helical domains of the protein. Adapted from reference [141].

The FliG monomer is completely helical and consists of three globular domains connected by two helices. Mutations were mapped that biased the direction of rotation, and the data suggested a conformational change from a closed conformation (anticlockwise) to an open conformation (clockwise) due to dissociation of helix$_{MC}$ from the ARM$_M$ domain (Fig. 3.9). Hydrodynamic modelling of the directional switching supports much of the experimental evidence.[142]

The Type III protein export system, associated with the flagella, also shows striking similarities to F_0F_1-ATP synthase, FliI is the ATPase of the export system and this protein shows strong homologies to the amino acid sequence of both the α and β subunits of F_0F_1-ATP synthase.[143] FliI works with FliH as a dimer and enables delivery of proteins to the export gate at the base of the C-ring in an ATP-dependent manner. However, this ATPase activity is not an absolute requirement for protein export, which instead appears to be dependent upon the proton flux[144,145] through the motor, while the ATPase activity appears to be involved in release of the FliH–FliI complex from the export gate.

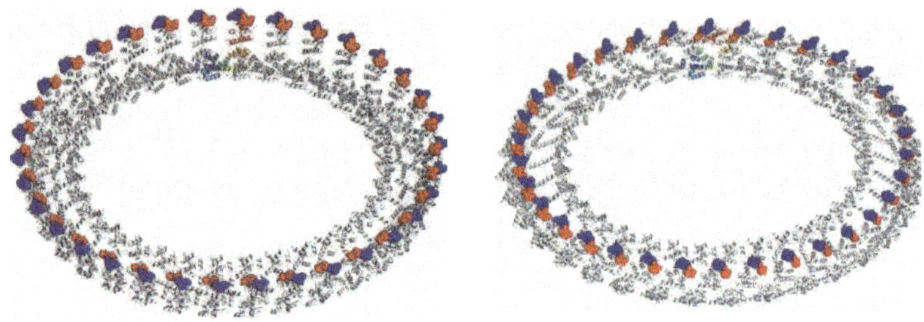

(a) (b)

Figure 3.9 Molecular basis of direction switching of the bacterial flagellum. Transition between the open (a) and closed (b) conformation of FliG produces clockwise or anticlockwise rotation, respectively. Reprinted by permission from Macmillan Publishers Ltd from reference [141]. Copyright 2010.

The assumption made in much of the work detailed above is that the structure of the flagella motor is static, once assembled. However, recent work has suggested that the flagella motor may be a much more dynamic system than originally thought,[137,146,147] and this has been investigated in a detailed manner using green fluorescent protein (GFP) gene fusions.[148] These GFP-FliN, GFP-FliM, and GFP-FliG fusions were incorporated into a functional flagella motor and then photobleached to remove their fluorescent output. Recovery of the fluorescence was then observed as freshly produced motor protein-fusions replaced the photobleached versions. The rate of this turnover was found to be 0.0007 s^{-1}, which equates to all of the FliN protein being replaced within tens of minutes. Therefore, this biological motor is a highly dynamic system.

Several methods have been employed to measure the forces generated by the flagella motor either by tethering the cell and measuring viscous drag,[149] or by tethering cells on latex beads held in an optical trap.[150–152] If the average of these measurements is used, it is possible to calculate the force per individual motor (MotA, MotB) component at ~25 pN.[124] Recent mathematical models of this motor activity have confirmed many of the observations and allowed predictions of step size and stator number.[153,154]

3.3 P-Type Ion Pumps

Ion pumps of this type are all ATPases, and the obvious connection between the F_0 section of ATP synthase and the bacterial flagella motor means it is appropriate to mention the other P-type ion pumps in this section of the book. They include Ca^{2+}-ATPase, Na^+/K^+-ATPase, and H^+-ATPase and are responsible for maintaining electrochemical gradients of these key cations.[155,156] There are five families — P_I–P_V, and the best studied systems belong to the P_{II} family (Ca^{2+}-ATPase from the P_{IIa}, Na^+/K^+-ATPase and H^+/K^+-ATPase from the P_{IIc} sub-family).[122]

The mechanism of action of the pumps involves auto-phosphorylation and auto-dephosphorylation at a conserved aspartate residue and the conserved TGES (Thr-Gly-Glu-Ser) motif that stabilises the transition states during phosphoryl-group transfer.[157] Delocalisation of the charge of the γ-P of ATP occurs through the carboxyl group of Asp^{351}, Lys^{654}, and the Mg^{2+} cofactor, and the subsequent release of the γ-P unlocks an inter-domain interface, allowing a transition from a high-energy to a low-energy state and leads to rotation of the A-domain by 120° with the TGES domain replacing ATP (c.f. F_0-ATP synthase).[156]

3.4 Bacteriophage DNA-Loading Motors

A viral genome is packaged into the viral head[158] of most bacteriophage (prokaryotic viruses) and many eukaryotic viruses (e.g., herpes simplex virus) using a DNA-loading motor,[159] a process that involves compacting the DNA into such a small space that the density of the DNA is similar to that obtained in crystalline structures.[160] The internal forces on the packaged DNA have been estimated at around 50 pN,[161] which are due to electrostatic repulsion due to the negatively charged backbone of the DNA and due to DNA bending against a persistence length for DNA of 50 nm.[162] The internal pressure created has been shown to be sufficient to be capable of "bursting" the viral head,[163,164] but the structure is maintained during the packaging process by an associated stabilisation protein (e.g., gpD from λ). This internal pressure provides sufficient force to drive partial, but not complete, ejection of the phage DNA during infection.

Primarily for historical reasons, the motor system used to package bacteriophage DNA has been known as the terminase[165] or portal protein. It comprises a cone-shaped dodecamer of subunits, arranged radially around a central channel through which the DNA can pass (Fig. 3.10). The terminase, or portal protein complex, is composed of two main proteins (Table 3.1): a small terminase protein, which binds to the bacteriophage concatameric DNA in a specific manner, near the site for cleavage (where the concatameric DNA is cut into genome-sized lengths — e.g., the 17 bp *pac* site for P22 bacteriophage); and a large terminase protein, which is responsible for this DNA cleavage and has an ATPase activity (that powers DNA translocation) and a docking domain that locates the vertex of the icosohedral prohead of the virus. Sequence alignments of the genes involved in production of these proteins suggest a common ancestor,[166–168] and this conservation is also apparent in herpes simplex virus (HSV1, Table 3.1).

The bacteriophage ϕ29 is unusual in that the gp3 protein is also a component of the DNA replication system and remains covalently attached to the viral DNA during packaging. The large terminase component, unusually, also includes a 174 nt RNA molecule (packaging RNA–pRNA), which enables docking of gp16 onto the portal protein.[169] The pRNA forms a penta- or hexameric ring structure, which has been shown through NMR studies[170] to be linked by internal base pairing. The pRNA is embedded within the portal vertex of the capsid, with a central channel for DNA to pass through and bridges two protein components (gp10 and gp16 — the large terminase protein). The pRNA binds the N-terminus of gp10[171,172] and allows assembly of gp16 to produce the functional motor.

Table 3.1 Proteins involved in the assembly of phage DNA packaging

	Small terminase component		Large terminase component	
Phage/virus	Gene product	Molecular mass (kDa)	Gene product	Molecular mass (kDa)
λ	gpNu1	20	gpA	73
T4	gp16	18	gp17	70
T3/T7	gp18	20	gp19	67
SPP1	gp1	21	gp2	49
P22	gp3	19	gp2	58
ϕ29	gp3	31	gp16	39
			pRNA	58
T5	?	?	gp155	49
HSV1	UL28	85	UL15	81

Source: Adapted from reference [165].

Viral assembly involves production of a prohead from the capsid proteins, into which the DNA will be packaged prior to motor protein removal and attachment of the viral tail. The motor is located at a unique position on the icosahedral capsid (Fig. 3.10), and one ATP molecule translocates 2 bp of DNA[173,174] and generates the highest known force for any biological molecular motor — 60 pN per molecule.[161,164,175] Essential to the assembly of the prohead is the presence of the portal protein, capsid proteins, and, in most cases, a scaffold protein that is removed in the mature virion.

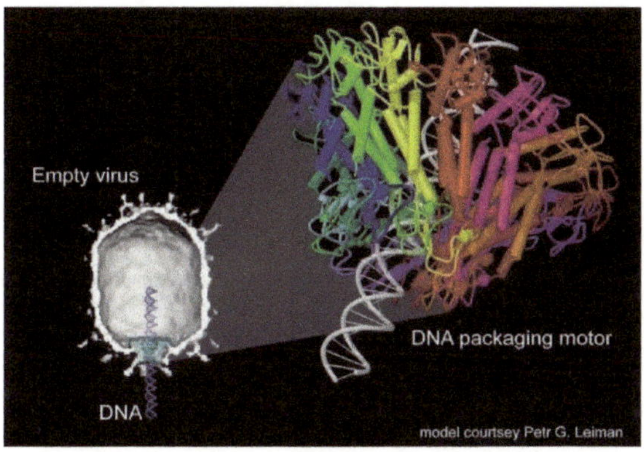

Figure 3.10 Location and arrangement of the viral-head packaging motor from Phi29. This computer illustration shows the arrangement of proteins that make up the bacteriophage Phi29 DNA-packaging motor. The tiny motor is positioned at the entrance of the viral shell and feeds viral DNA into the shell as new viruses are assembled in a host cell (Purdue News Service Graphic courtesy of Clark Gedney and Petr G. Leiman).

3.4.1 Structural Features of DNA-Packaging Motors

A crystal structure exists for ϕ29 and SPP1 packaging motors[176,177] and cyro-EM images exist for ϕ29,[176] SPP1,[178] and T4.[179] These images show that the wider end of a cone-like structure (Fig. 3.10) lies within the capsid with the narrow end protruding out. Figure 3.11 shows the structure of an individual subunit of the portal protein and several domains are apparent:

1. The "stalk" structure, which protrudes from the prohead of the virus and allows assembly of the motor complex.

2. The "stem", which has two long alpha-helices that, in the oligomeric assembly, provide a central cylinder walled by 24 of these helices.
3. The "wing" and the "crown", which are domains located internally within the prohead of the virus.

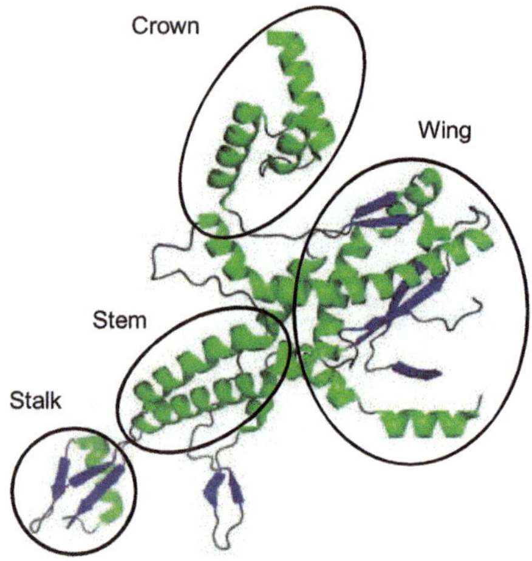

Figure 3.11 Structure of the portal protein of the packaging motor. Details showing a ribbon structural representation of the bacteriophage packaging motor monomer unit that oligomerises to produce the functional motor. Adapted from reference [165].

There are significant regions of conservation between the protein sequences of various viral DNA-packaging proteins, despite their overall lack of similarity (Fig. 3.12), and these have allowed the identification, through mutagenesis studies, of a number of important motifs.[180,181]

The small terminase protein of bacteriophage lambda shows an N-terminal helix-turn-helix motif, typical of many DNA-binding proteins, which has been shown to contact the *cosB* DNA sequence,[182,183] producing a bend in the DNA. This helix-turn-helix motif seems to be a common feature of these proteins.

Small terminase proteins oligomerise in solution producing oligomers with 8–10 subunits, and this process appears to involve conserved coiled-coil motifs that are critical for function and form parallel arrays to create the oligomers.[184–186] Amino acids 18–35 of the lambda small terminase protein (gpNu1) are involved in DNA binding, but there is no classic Walker A or B site and only weak ATP binding is observed. The C-terminus of the small terminase protein interacts with the N-terminus of the large terminase protein, and this interaction is required to bring the cleavage-domain of the large terminase protein into contact with the DNA bound by the small terminase protein.[187] However, there is no direct knowledge of the amino acids involved, and this (especially for the T4 bacteriophage system) appears to reflect a dynamic interaction. Specificity components within the small and large terminases of T4 packaging motors have recently been identified, which should improve the understanding of how DNA packaging occurs.[188]

```
T4        NH2(137)QLRDYQRDML( 9)VCNLSRQLGKTTVVA( 75)NSFAMIYIDECAFI( 20)KIIITTTPNGL(319)COOH2
RB49      NH2(135)QLRDYQKDML( 9)AHKLSRQLGKTTAVA( 75)NSFSMIYIDECAFI( 20)KMINTTTPNGL(318)COOH2
T7        NH2( 33)VPTKCQKDML(11)ILQAFRGIGKSFITC( 82)SRADIIIADDVEIP( 29)RVIYLGTPQTE(381)COOH2
T5        NH2( 33)TPNGPQIAII(11)TACVSRRVGKSFIAY( 73)RSYDFIIFDEAAIS( 20)KALFISTPRGG(248)COOH2
λ         NH2( 41)KESAYQEGRW(21)NVVKSARVGYSKMLL( 82)KSVDVAGYDELAAF( 24)KSIRGSTPKVR(423)COOH2
SPP1      NH2(  7)EKFTPHFLEV(12)VLKGGRGSAKSTHIA( 81)FPVAGMWIEELAEF( 24)IFFYSYNPPKR(248)COOH2
P22       NH2( 37)APYSKQREFI( 9)CFMAGNQLGKSFTGA(123)DTIHGVWFDEEPPY( 15)FSILTFTPLMG(265)COOH2
ϕ29       NH2(  4)LFYNPQKMLS( 6)FVIGARGIGKSYAMK( 72)PNVSTIVFDEFIRE( 28)RCICLSNAVSV(170)COOH2
HK97      NH2( 27)RLDPFQKDFI(13)ILSIARKNGKTGLIA( 79)LSPILAILDETGQV( 21)LLIVISTQAAN(284)COOH2
HSV-1     NH2(221)GDHAEQVNTF(23)VFLVPRRHGKTWFLV( 78)QDFNLLFVDEANFI( 17)KIIFVSSTNTG(350)COOH2
```

Secondary structure 

SF2 Helicases

```
RAD3      NH2( 15)KITPEQYNYM(13)ILEMPSGTGKTVSLL(173)SKDSIVIFDEAHNI(215)VIITSGTISPL(312)COOH2
NS3       NH2(181)PVFTDNSSPP( 9)HLHAPTGSGKSTKVP( 66)GAYDIIICDECHST( 22)VVLATATPPGS(303)COOH2
EIF4A     NH2( 43)EPSAIQQRAI( 9)LAQAQSGTGKTGTFS( 84)DKIKMFILDEADEM( 21)VVLASATMPND(188)COOH2
```

Type I restriction endonucleases

```
EcoKI     NH2(461)GLRYYQEDAV(11)LLAMATGTGKTRTAI( 83)ARYDCIVVDEAHRG( 34)KIALTATPALH(546)COOH2
EcoR124I  NH2(270)VMPRYQEDAV(11)LIWHTTGSGKTLTSF( 80)NQQVVFIFDECHRS( 17)QFGFTGTPIFP(597)COOH2
KpnAI     NH2(293)CCRYPQYYAG(20)TYFGATGCGKSYTMQ( 92)RSNIICISDEAHRS( 35)YVGFTGTPIDA(621)COOH2
```

Type III restriction endonucleases

```
LlaFI     NH2( 43)AFRRFQMQMD(16)LFNMATGSGKTMVMA( 99)DEDIVILGDEAHHF( 29)LLEFSATINMD(634)COOH2
EcoPI     NH2( 55)NIKKVQELNG(15)DVSMETGTGKTYTYT(122)AVKPFIIIDEPHKF( 17)IIRYGATFSEG(701)COOH2
StyLTI    NH2( 24)GIDHAQADHN(17)DVKMETGTGKTYVYT(123)MTRPVVIIDEPHRF( 17)IVRFGATFPDI(759)COOH2
```

Protein translocases

```
FtsH      NH2(176)NPSRFHEMGA( 6)LLVGPPGVGKTHLAR( 39)HAPCIVFIDEIDAV( 35)VVMAATNRPDI(319)COOH2
CLPA      NH2(210)TTNLNQLARV(28)LLVGESGVGKTAIAE( 51)DTNSILFIDEIHTI( 35)TYQEFSNIFEK(435)COOH2
SecA      NH2( 79)FPFKVQLMGV( 7)IAEMKTGEGKTLTST( 95)DEVDSILIDEARTP(146)LAGMTGTAKTE(464)COOH2
```

Adenine Binding | Walker A | Walker B | Catalytic carboxylate | C-motif

Figure 3.12 Regions of similarity between DNA-packaging motors. Adapted from reference [165].

The large terminase protein has two domains: an N-terminal ATPase required for DNA translocation and a C-terminal nuclease domain that must be physically connected for translocation to occur. Alignments show one Walker A ATP binding site in the N-terminus of the large terminase protein.[168] There is also a Walker B motif, which will tolerate some substitutions as long as both the hydrophobicity and β-strand structure are preserved; other substitutions often bind ATP, but do not function.[189] A carboxylate group (usually Glu) located between the Walker A and B motifs[168] acts as a base, through a water molecule, leading to ATP hydrolysis (c.f. ATP synthase), and this is located adjacent to the Asp residue of the Walker B motif (c.f. helicases and RecA).[190,191] In fact, most helicases have an ATP coupling motif,[192] which is also found in the large terminase proteins (Motif III) that forms a network of hydrogen bonds between the γ-phosphate of ATP and DNA and is required for movement. Mutations within Motif III of the T4 packaging motor are affected in turnover and will bind a single ATP molecule, hydrolyse it, but lose translocation activity[167]; while similar mutations in the lambda large terminase protein lead to only a part of the DNA being packaged.[193] There is also a conserved two amino acid sequence in the large terminase protein (YQ), situated ~15 amino acids upstream of the Walker A motif, that produce aromatic stacking interactions with an adenine in the DNA, which is required for DNA packaging. Finally, in a manner that is analogous to the "finger" of ATP synthase, there is also a mechanism within the large terminase protein for release of Pi (Arg + Glu). A packaging model has been proposed[194] (Fig. 3.13) in which Arg^{162} of the T4 motor is a key component for "firing" movement, with the N-terminal ATPase sub-domain I being the "engine" and the C-terminal domain acting as a wheel.[195] The N-terminal sub-domain II is a gear box that transmits energy from ATP hydrolysis to the C-terminal domain where translocation is carried out. This results in a compression–relaxation activity, which causes portal rotation and is coupled to DNA translocation.[165] In the pre-translocation complex, the flexible linker between the N- and C-terminal domains is extended. Binding of DNA (by the C-terminal domain) and ATP to the N-terminal domain of the same gp17 molecule produces conformational changes, which are transmitted to sub-domain II resulting in insertion of Arg^{162} into the ATPase active centre to trigger ATP hydrolysis. Release of ADP and Pi causes sub-domain II to rotate slightly, pulling the flexible linker and aligning charged pairs of amino acids between the N- and C-terminal domains. This process translocates 2 bp of DNA. The loss of negative charge associated with dissociation of ADP and Pi attenuates the electrostatic interactions between the N- and C-domains, whereupon sub-domain II rotates back, charged pairs between sub-domain I and the C-terminal domain are misaligned and return the C-terminal domain to a "relaxed" state. Translocation of 2 bp positions the neighbouring C-terminal domain in an optimal position for DNA binding, and the process can repeat for the next motor protein, thus DNA is handed over to each neighbouring C-terminal domain (see movies depicting this motion at http://www.cell.com/supplemental/S0092-8674(08)01447-5).

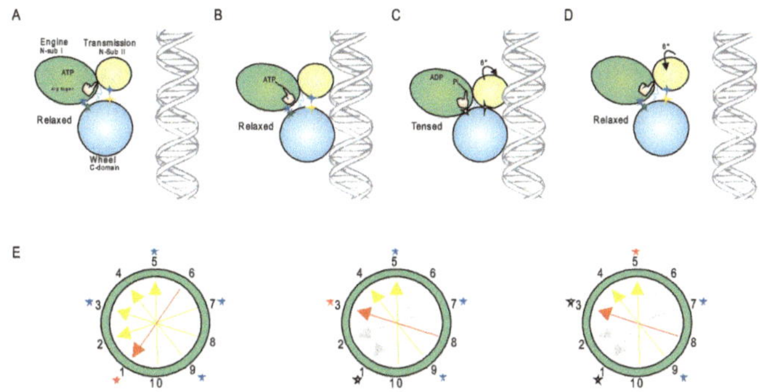

Figure 3.13 Proposed DNA-packaging mechanism. Panels (A–D) show the conformational changes associated with DNA binding and movement for a single gp17 molecule. The gp17 N-terminal sub-domain I, sub-domain II, and C-terminal domain are represented as green, yellow, and cyan ovals, respectively. The five-pointed stars show the charge interactions between the N-terminal sub-domain I and the C-terminal domain. The four-pointed stars show the charge interaction between the N-terminal sub-domain II and the C-terminal domain. The flexible linker between N- and C-terminal domains is represented by a wiggly cyan line. (E) This panel relates to the synchronisation of the five gp17 molecules located around the vertex of the procapsid. Successive DNA base pairs are indicated by yellow arrows outside the procapsid, red entering the procapsid, and white inside the procapsid. The surrounding five gp17 molecules are shown as stars. The red star represents gp17-ATPase-hydrolysing ATP, the blue star represents ATPase ready to hydrolyse ATP, and the white star represents ATPase that has already hydrolysed ATP. Adapted from reference [194].

This motor process requires information exchange between the neighbouring gp17 molecules in order to synchronise ATP hydrolysis. This process must be through the DNA and seems to be related to the 10-fold symmetry of the DNA and the 5-fold symmetry of the gp17 molecules in the oligomer structure (i.e., it is the relationship of the DNA structure to a specific gp17 monomer that determines the timing of ATP hydrolysis). Therefore, four gp17 molecules are simultaneously in a "relaxed" state, while one molecule is in the "tensed" state.

Phylogenetic studies and sequence alignments (Fig. 3.12) suggest a close relationship between these motors and the superfamily 2 helicases,[167] including Type I and III restriction-modification enzymes, and the motor activity reflects this connection.

Single molecule studies of the bacteriophage DNA-loading motors have involved the use of an optical tweezer system[161,163,164,196–200] in which a bacteriophage head is attached and immobilised on a large bead coated with antibody, in such a way that it will still load viral DNA, and the DNA is "pulled" back from the loading motor by a magnetic tweezer system (Fig. 3.14). This has allowed the step size of the motor to be determined as 2 bp per ATP molecule hydrolysed,[173,174] which fits with the 10-fold rotational symmetry of DNA. The stalling force, determined by "pulling back" with the optical tweezer set-up, as mentioned earlier, was found to be very high (as expected to enable packaging against the forces generated within the prohead) at 60 pN,[161,163] which is 10x that of kinesin (Fig. 3.14) and 20–25x that of myosin.[201]

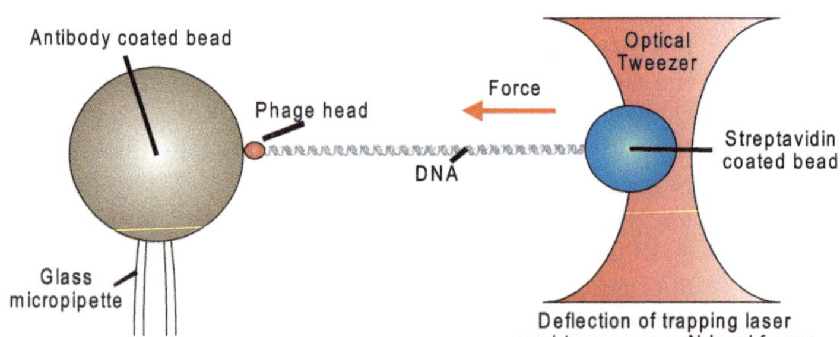

Figure 3.14 The use of an optical tweezer set-up to study bacteriophage DNA-loading motor. Experimental set-up for a single-motor experiment, using an optical tweezer set-up, to measure the forces produced by the DNA-packaging motor. The phage head is attached to an immobilising bead, using an antibody coating on the bead, which is held in position by a micropipette system. DNA is attached to a streptavidin-coated polystyrene bead using a biotin incorporated into the end of the DNA and pulled using an optical trap set-up consisting of a tightly focussed laser light (pink). Adapted from http://physics.ucsd.edu/~des/Viral_DNA_Packaging.html.

The motors show high processivity, primarily due to the fact they wrap around the translocated DNA, and high speeds of translocation.[164] Interestingly, the rate of translocation decreases as the head is filled (about three-fold in lambda), presumably because of the counter force associated with packaging of the DNA into the prohead, and the velocity appears to be related to genome size (Table 3.2). In most systems, approximately 30% of the energy available from ATP hydrolysis is converted into DNA translocation, with much of the rest produced as heat, which may provide an entropic contribution to offset the energy required for DNA packaging.

Table 3.2 Relationship between rate of DNA translocation and genome size for bacteriophage DNA-loading motors

Bacteriophage	Genome size (kbp)	Rate of translocation (bp s^{-1})
ϕ29	20	100–150
λ	48.5	600
T4	170	~700

Source: Adapted from references [163, 164].

3.5 Potential for Using Biological Rotary Motors in Bionanotechnology

The rotary biological molecular motors described above have become model systems for the design of some of the previously described chemical motors, but powerful rotary motors are also perfect systems for the design of nanoscale and microscale devices (see Chapter 5). However, as detailed in Table 3.3, the motors most likely to find uses in bionanotechnology are ATP synthase and flagella motor. ATP synthase has both potential as a propeller for directional motion, but also as a source of ATP as a step between light-generated proton generation and motor-driven electricity generation. While the bacterial flagella motor has the capability of rotating microscale objects attached through a membrane structure.

Table 3.3 Summary of available rotary biological motor proteins

Name of motor	Primary function	Source of protein	Most likely use in bionanotechnology	Reference
ATP synthase	Generation of ATP.	Boris Feniouk JST ICORP ATP, National Museum of Emerging Science and Innovation 2-41 Aomi, Koto-ku, Tokyo 135 0064, Japan.	Propeller. Biological generation of electricity.	202
E. coli flagellum	Rotation of bacterial flagellum.	Richard M. Berry The Randall Centre, King's College London SE1 1UL, UK	Rotating microscale objects.	

Chapter 4

Biological Motors That Produce Linear Motion

4.1 Kinesins

When considering biological molecular motors that may have applications in bionanotechnology, one of the first types that spring to mind are kinesin superfamily proteins (KIFs). Since their first identification during studies of squid axonal transport,[203] kinesins have been identified and purified from a diverse range of sources, which has allowed for the characterisation of a number of classes.

The kinesin superfamily constitutes motor proteins that travel along microtubules and conduct a range of functions, including transport of cargo (vesicles, organelles, chromosomes, protein complexes, and ribonucleoproteins), and they are also known to be involved in microtubule dynamics.[204] Motility studies have shown that kinesin moves in a unidirectional manner along microtubule tracks and that this movement corresponds with movement *in vivo*, which is away from the microtubule organising centre towards the cell periphery. Single-molecule work[205] has shown kinesin dimers move in a highly processive manner from one α/β tubulin subunit to the next in a series of ATP-dependent steps of 8 nm each. The rate of movement is approximately 100 steps per second, and each step, following hydrolysis of a single ATP molecule, produces approximately 5–7 pN of force.[206] Kinesin produces this movement by a "hand-over-hand" mechanism involving several steps that

Molecular Motors in Bionanotechnology
James Youell and Keith Firman

ISBN 978-981-4267-02-1 (Hardcover), 978-981-4303-09-5 (eBook)
www.panstanford.com

depend upon the presence or absence of ADP, or ATP. The key to the movement is a structural change induced by ATP binding that allows the free head to locate another binding domain on the microtubulin via diffusion. This process is essentially an ATP-driven means of directed communication between the subunits (head units).

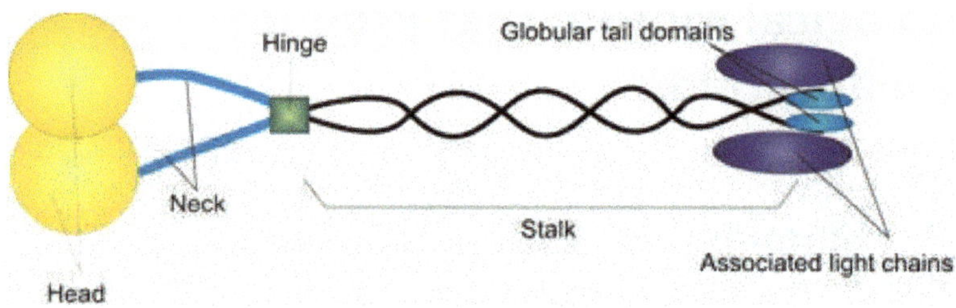

Figure 4.1 Kinesin complex. Illustration of the conventional kinesin heavy chain dimer showing the proposed domain architecture.

In general, the majority of KIFs have a globular motor domain, incorporating an ATPase site and a microtubule binding motif, termed the "head" (Fig. 4.1). The hydrolysis of ATP by the head results in the translocation of the kinesin molecules along the microtubule substrate with directionality (normally towards the plus end of microtubules). The motor domain itself is relatively well conserved amongst the KIFs, but other regions are variable reflecting the varied roles and associated proteins and light chains of the different kinesins. Generally the kinesin heavy chains dimerise via their stalk regions, although there are examples of monomeric and heterodimeric kinesin complexes. Cargoes are found bound through the highly variable tail regions, either directly or indirectly through light chain or associated proteins.[207–209]

4.1.1 Nomenclature

The original nomenclature for KIFs (pre 2004) was dependent on various factors, including the order of discovery, sequence analysis, overall structure, and positioning of the motor core. Leaders in the field, in order to minimise the confusion brought about by this seemingly haphazard scheme, standardised the nomenclature in 2004. The individual sequence names remained unchanged, but the family names were all assigned the name "kinesin" and members were grouped into families based on a systematic phylogenetic analysis and, therefore, structural relatedness[210] (see Table 4.1).

Table 4.1 Summary of kinesin types

		Other names for this group of sequences			
Standardised names	**Example sequences**	**Endow *et al.***	**Hirokawa**	**Lawrence *et al.***	**Other names**
Kinesin-1	KHC KIF5A KHC K7	KHC	N-I	Kinesin-I	Conventional
Kinesin-2	KRP85 KRP95 KIF3A KIF3B FLA10	KRP85/95	N-IV	Kinesin-II	Hetero-trimeric
Kinesin-3	KIF1A UNC104 KIN Unc104	Unc-104/Kif1	N-III	Unc104	Monomeric
Kinesin-4	KIF4 CHRKIN XKLP1 F11C1.80 AY224568 K8	Chromokinesin/ Kif4	N-V	Chromokinesin (upper clade)	
Kinesin-5	KIF11 Eg5 BIMC KRP125	BimC	N-II	BIMC	Eg5 Bipolar Tetrameric
Kinesin-6	MKLP1 KIF20A K12	MKLP1	N-VI	MKLP (lower clade)	Cho1
Kinesin-7	CENP-E KIF10 KIP2p AB028470	CENP-E	N-VII	CENP-E	
Kinesin-8	KIF19A KIP3p T9C5.240	Kip3	N-IIX	Kip3	
Kinesin-9	KIF8 KLP1			MKLP (middle clade)	
Kinesin-10	Nod KID KIF22 T1E22.130			Chromokinesin (lower clade)	Kid

(Continued)

Table 4.1 *(Continued)*

		Other names for this group of sequences			
Standardised names	**Example sequences**	**Endow *et al.***	**Hirokawa**	**Lawrence *et al.***	**Other names**
Kinesin-11	VAB-8 KIF26A SMY1p			Divergent Kinesin-I	
Kinesin-12	KIF15 Xklp2 PaKRP1 3g23670 JGI_3356			MKLP (upper clade)	
Kinesin-13	MCAK KIF2 3g16060 DSK1	MmKif2, MCAK/Kif2	M	I-Type	KinI
Kinesin-14	NCD KIFC1 KAR3p KATA KCBP	C-terminal motor	C	C-type	
Orphan kinesins		Orphans		Orphans	Ungrouped

Source: Adapted from bibliography reference [210].

4.1.2 General Features of the Kinesin Superfamily Proteins (KIFs)

Most known kinesin molecules have a "catalytic core", which is a ~360 amino acids globular domain responsible for ATP hydrolysis and microtubule binding.[207–209,211] Almost all of the members of the kinesin superfamily form a long filamentous structure, with one end comprising the globular domain and the other a fan-shaped structure responsible for binding kinesin light chains. Many of the structural features of kinesins are named through analogy with the features of kinesin-1. Therefore, in the majority of kinesins, the catalytic domain resides at the N-terminal end of the peptide, designated the "head" region, followed by the "stalk" region, which joins it to the "tail" region at the C-terminal end of the molecule.[207–209,211]

Research has shown that the "head" domain is the main catalytic core of the molecule causing movement of the kinesin along microtubules,[208,212,213] powered by the hydrolysis of ATP, whereas the "tail" region is important for cargo binding or interaction with other subunits. The cargo bound obviously depends upon the role of the kinesin in question and cargo molecules such as proteins, lipids, or nucleic acids have all been observed.[208,214,215]

Between the "head" and the "stalk" region lies the "neck" region of the kinesin molecule. The "neck" region often has features that are specific to certain kinesin families and has been shown to be essential for force generation, activity regulation, and control of directional movement.[216–220]

In what could be considered atypical kinesins, the motor domain may reside in the middle or the C-terminus of the peptide (and indeed, as has been alluded to previously, this formed the basis for one form of nomenclature proposed prior to that of Lawrence *et al.*[210]) and this, in turn, appears to determine the direction of translocation and the role of the specific kinesin.

4.1.3 Kinesin Families

Based on a large systematic phylogenetic analysis across over 640 KIF sequences, 623 catalytic core sequences allowed the determination of 14 distinct KIF families[221] (illustration domain layouts are shown in Fig. 4.2–4.5), which supported the standardised nomenclature system.[210]

4.1.3.1 Kinesin-1 family

The first of the KIF to be discovered was kinesin or kinesin heavy chain (KHC), and it is grouped within this family (Fig. 4.2). The defining feature of the kinesin-1 family is the family specific β-sheet and the highly conserved coiled-coil stalk.[221] The kinesin-1 family reportedly work as organelle transporters[222] and are involved in nuclear movement[223] and encompasses three subfamilies: animal KHCs, fungal/plant KHCs, and atypical KHCs.[221]

The tetrameric animal KHCs (e.g., KIF5A, KIF5B, etc.) have both a cargo binding domain (CBD)[224] and also a light chain binding domain (LCBD) in the stalk region. The LCBD binds to an adaptor-functioning kinesin light chain (KLC),[225] which can also allow binding of cargo at the C-terminus.[226]

Fungal and plant KHCs are homodimers and lack LCBDs but are otherwise thought to function in a similar fashion. Atypical

KHCs lack both CBDs and LCBDs but instead comprise either a trans-membrane domain or an armadillo (ARM) domain (a region of protein–protein interaction mediated through a 42–45 amino acid repeat), which has been suggested as an alternative site for cargo or adaptor binding in the absence of adaptor proteins. It is thought that these atypical KHCs may potentially fulfil the roles of kinesin-2 family members, which are absent in higher plants.[221]

4.1.3.2 Kinesin-2 family

The kinesin-2 family were originally identified as the N-kinesins and have been further divided into three subfamilies: KIF3A, KIF3B/C, and KIF17 subfamilies (kinesin-2 family, Fig. 4.2).

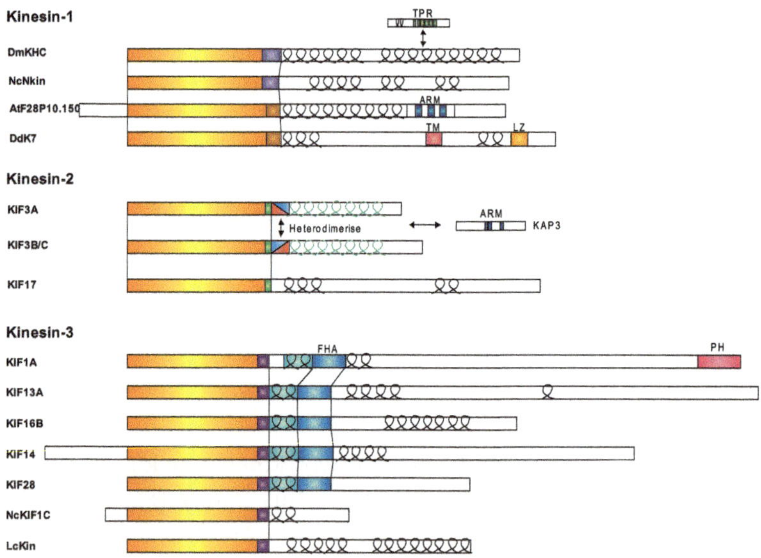

Figure 4.2 Structure of kinesin families 1 to 3. Illustration of the structures and domains of representative members of each kinesin superfamily protein (KIF) family and subfamily. The family/subfamily specific N-terminal accessory domains (where present), motor domains, neck domains, and C-terminal stalk regions are depicted as discussed in the relevant sections of the text. Adapted from Miki *et al.*[221]

The KIF3A subfamily sequence contains clusters of positively and negatively charged residues immediately after the neck β-sheet following the catalytic core. The KIF3B/C subfamily has a similar structure to that of the KIF3A subfamily; however, the alternating clusters of positively and negatively charged sequences are in opposition to those of the KIF3A subfamily, which is thought to give rise to hetero-dimerisation between these subfamilies.[227] By contrast, however, the KIF17 subfamily members are reported to homo-dimerise.[228]

Members of all the subfamilies have been discovered to participate in organelle transport,[228] intra-flagella transport,[229] and spermatogenesis.[230]

Research has not yet identified any kinesin-2 members from fungi or higher plants. It has been proposed that the lack of members in plants could be explained by compensation involving kinesin-1 members[221] as mentioned above. KIF3s are known to closely associate with KAP3, an adapter molecule containing ARM repeats.[229,231,232] Atypical kinesin-1 plant KHC members, which contain these ARM repeats, could potentially function as intrinsic adapter molecules at the site of adaptor protein binding.

4.1.3.3 Kinesin-3 family

The original classification of the founding member for this family was the N-III kinesin.[207] The major characteristics of this family are the conformation of the β-sheet and helical region, followed by a forkhead-associated (FHA) domain[233] (kinesin-3 family, Fig. 4.2). The FHA is a site of protein–protein interaction with phosphor-proteins, which could mediate binding[234] and it has also been indicated as being involved in intramolecular interactions for self-controlling motor activity.[235]

The kinesin-3 family is composed of five subfamilies — KIF1, KIF13, KIF14, KIF16, and KIF28 subfamilies — and an uncategorised group — the NcKIF1C/Klp7 group. Similarly, to the kinesin-2 family, the members of these subfamilies are found in animals, fungi, and multiple protozoa, but not in higher plants.[221]

All members of the subfamilies as yet characterised have been discovered to be involved in organelle transport,[236–238] and their members are reportedly either monomeric[237,239] or homodimeric.[240]

4.1.3.4 Kinesin-4 family

The kinesin-4 family had also been known as the N-V,[207,241] KIN N-Chromo,[242] chromokinesin/KIF4,[243] and the chromokinesin[244] family before the more recent nomenclature system. The conservation of the neck β-sheet and coiled-coil stalk region of the kinesin-4 family members is high and members have been split into five subfamilies: the KIF4, KIF7, KIF21, KIF27, and NcKIF21A subfamilies (kinesin-4 family, Fig. 4.3).

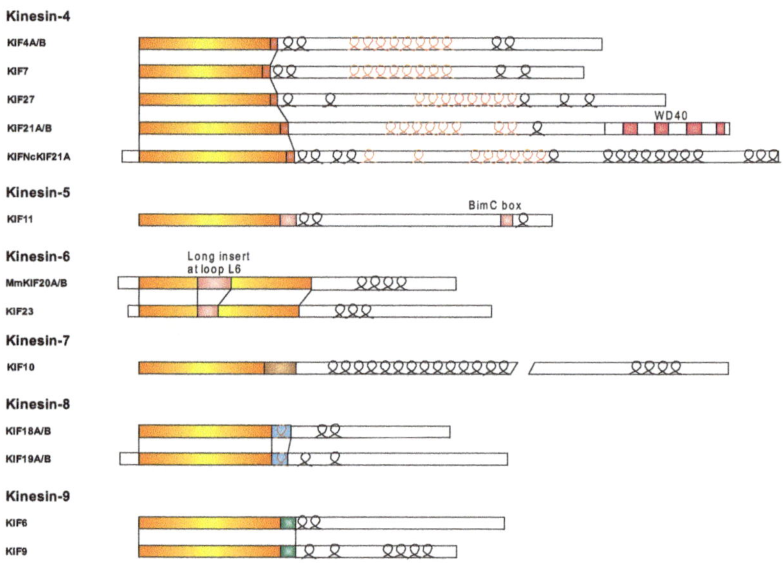

Figure 4.3 Structure of the kinesin families 4 to 9. As in Fig. 4.2.

The family members are thought to function in processes ranging from organelle transport[245] to chromosome movement[246,247] and spindle formation.[248] Similar to the kinesin-2 and kinesin-3 families, the kinesin-4 family has more family members in higher vertebrates, fewer members in lower vertebrates, and a few, if any, in plants.[221]

4.1.3.5 Kinesin-5 family

Similar to the kinesin-4 family, the kinesin-5 family was numbered based on the animal homolog Eg5.[249,250] The family has also been known as the N-II,[207,241] KIN N-Bipolar,[242] and BimC[243,244] prior to the agreed nomenclature, and family members exhibit the highest conservation amongst the kinesin families, with little sequence divergence in comparison to other families.[221] Due to this, the family cannot be further subdivided (kinesin-5 family, Fig. 4.3).

Members of this family have been characterised from mammals, yeast, and higher plants, and the overall structure is similar for all members, with a family-specific neck downstream of the catalytic core and a BimC box domain at the C-terminus.[251] The BimC box domain is phosphorylated by the p34/Cdc2 kinase,[250] localised to microtubules[252] and is linked with competence for cell division.[253] The members are reported to form homotetramers[254] and have been shown to be mitotic motors required for the formation of a bipolar spindle.[255]

4.1.3.6 Kinesin-6 family

The previous name for this family was N-VI[207,241] (also MKLP1[243] and MKLP[244]), and the numbering was continued for the formation of the kinesin-6 family. The neck region is more loosely conserved in this family; a more important characteristic is modification of the catalytic core in the form of a long insert in loop L6.[221] The loop is orientated on the side of the molecule opposite to the microtubule-binding surface,[256] and it is thought to possibly be involved in a mechanism for motor activity regulation as it is located on the cytoplasmic side of the molecule, potentially allowing access to cytosolic regulatory factors (kinesin-6 family, Fig. 4.3).

The kinesin-6 family is subdivided into two subfamilies: the KIF20 and KIF23 subfamilies. Members of the KIF20 discovered up to now include KIFs from fungi, animals, and a slime mould. KIF23 constitutes KIFs from animal sources alone. Neither of the two subfamilies is known to have plant KIF members.[221] The role of the members of the kinesin-6 family appears to be in cytokinesis[257,258] or microtubule transport.[259]

4.1.3.7 Kinesin-7 family

The numbering of this family has been retained from previous nomenclature, N-VII,[207,241] but has also been previously named as the CEMP-E[243,244] family. Members possess a long neck region specific to this family, and for those members from animals and plants, this neck region appears to be longer with an abundance of coiled-coil regions, in comparison to other isoforms that tend to be shorter with less coiled-coil regions (kinesin-7 family, Fig. 4.3). There appears to be a greater degree of conservation in animal kinesins, whereas plant kinesins show a greater diversity in sequence.[221] The members of the kinesin-7 family have been shown to have roles in mitosis, including microtubule capture at kinetochores[260] and nuclear migration.[261] The members of this family are widely conserved in most species; however, an abundance of sequences conforming to this family are found in plants suggesting diverse roles in higher plants, possibly including microtubule capture at phragmoplasts, the preprophase band, or cortical arrays.[221]

4.1.3.8 Kinesin-8 family

Member of the kinesin-8 (previously known as N-VIII[207,241] or KIP3[243,244]) family are characterised by their helical family-specific neck.[262] The kinesin-8 family is further divided into two subfamilies: an animal-specific subfamily containing KIF18, and a second subfamily that includes KIF19[221] (kinesin-8 family, Fig. 4.3).

Functions of the kinesin-8 family vary from nuclear migration[263] and spindle positioning,[264] to mitochondrial transport[265] and chromosome segregation[266] and alignment[267] during mitosis.

4.1.3.9 Kinesin-9 family

The naming of the kinesin-9 family is derived from one of its members KIF9. It is a novel family with a conserved family specific neck motif adjacent to, but downstream of the catalytic core. The catalytic core sequences additionally show some conservation of sequence between family members (kinesin-9 family, Fig. 4.3). The kinesin-9 family is split into two subfamilies comprising members from mammalians and protozoans, KIF9A, and KIF9B, and interestingly no members of the family have been found in invertebrates, fungi, or higher plants,[221] which has led researchers to hypothesise a role in cilia and flagella.[268]

In trypanosome flagellum, the function of members differs between the two subfamilies,[269] with KIF9A members required for flagellum motility[270,271] and implicated in dynein regulation,[272] whereas the KIF9B members appear to be essential for the assembly of the paraflagellar rod, where it is found associated with the basal body.[269]

4.1.3.10 Kinesin-10 family

The kinesin-10 family was first proposed by Miki *et al.*[221] and contains the KIF22, KID, and Nod proteins. Identification of members was primarily based on the C-terminal tail, where a family specific helix-hairpin-helix (HhH) is present. The neck sequence of this kinesin family is less well conserved, in comparison to other families (kinesin-10 family, Fig. 4.4). This has meant that previously members of the family have not been grouped systematically as analysis of sequence information has mainly been through comparison of neck and catalytic core sequences.[221]

KIFs of the kinesin-10 family are reported to function in cell division, especially chromosomal movements, which potentially could require the presence of the HhH motif.[273,274]

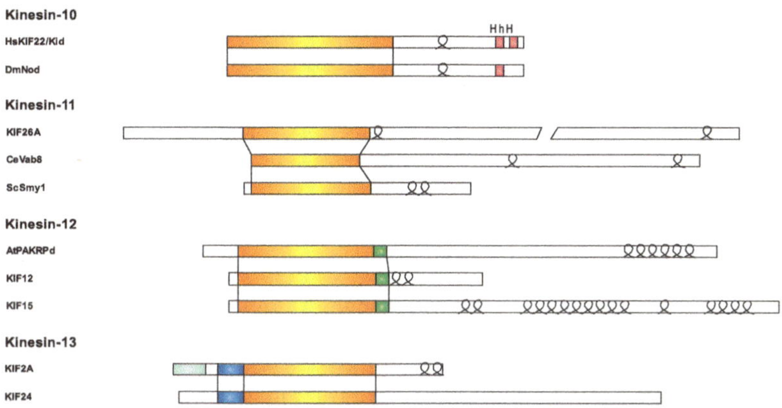

Figure 4.4 Structure of the kinesin families 10 to 13. As in Fig. 4.2.

4.1.3.11 Kinesin-11 family

This kinesin family encompasses those KIFs that contain an abnormal non-conserved catalytic core and has also been referred to as the N-XI family.[207,241] The founding member of the family is SmyI, which interestingly does not bind microtubules probably due to the unusual catalytic core.[275] There is no conservation of a family specific neck sequence, and there does not appear to be any subfamilies[221] (kinesin-11 family, Fig. 4.4). Kinesin-11 KIFs have been shown to be involved in signal transduction[276–278] and may well have no motility due to the structure of the catalytic core.[279]

4.1.3.12 Kinesin-12 family

The kinesin-12 family is numbered based on KIF12. The family comprises a combination of families formerly known as N-IX, N-X,[207,241] and AtPAKRP1.[243] The unification of these former families into the kinesin-12 family was justified by amino acid sequence homology that identified abundant coiled-coil sequence in the C-terminus of related proteins.

Three families have been identified that share a common neck β-sheet, with the KIF12 and KIF15 subfamilies sharing weak homology in the C-terminal coiled-coil region, which gave rise to the prediction of a common function between family members[221] (kinesin-12 family, Fig. 4.4).

In higher plants, members of the kinesin-12 family have been shown to transport organelles,[280] are essential for cell plate formation, act as dynamic linkers of the plus ends of antiparallel microtubules in the phragmoplast,[281] and essential for the spatial control of cytokinesis.[282]

Other members of this family have been shown to be essential for centrosome separation during mitosis[283] and even neuronal development.[284]

4.1.3.13 Kinesin-13 family

Members of what was to become members of the kinesin-13 family had catalytic cores in the middle of the molecule and, therefore, were originally referred to as the M Kinesin family[241] (middle-type motor) or KIN-I[242] (internal-type motor). This should not have been the case as numerous other KIFs from different families also show centralised catalytic domains and additionally some members of the kinesin-13 family have catalytic cores in the N-terminus. Also the term KinI resulted in a lot of confusion as it was often mistaken for kinesin-I, which refers to the kinesin-1 family.

The members of the kinesin-13 family can be characterised by their positively charged helical neck upstream of the catalytic core and can be divided into subfamilies: the animal-specific KIF2 and the KIF24 subfamily encompassing the remainder of the members of the family. The KIF2 subfamily has a catalytic core at the centre of the molecule whereas the KIF24 members have the catalytic core at the N-terminus[221] (kinesin-13 family, Fig. 4.4).

The kinesin-13 KIFs are involved in intracellular transport[285] and have been reported to have a microtubule-depolymerising activity,[286,287] which is an important factor in normal chromosome movement.[288] Mitotic centromere-associated kinesin (MCAK) has been found throughout the cell but is found at higher levels at the centromeres, kinetochores, and spindle poles,[289] and it has been shown that MCAK is essential for proper kinetochore–microtubule attachment during spindle formation.[290]

4.1.3.14 Kinesin-14 family

The kinesin-14 family is unique amongst the KIF families, as the protein is effectively reversed, with the neck sequences located N-terminally to the catalytic core, with members exhibiting a reversed motility due to the reversal in the direction of the motor domain.[291] The family is split into two subfamilies: the kinesin-14A subfamily and the kinesin-14B subfamily[221] (kinesin-14 family, Fig. 4.5).

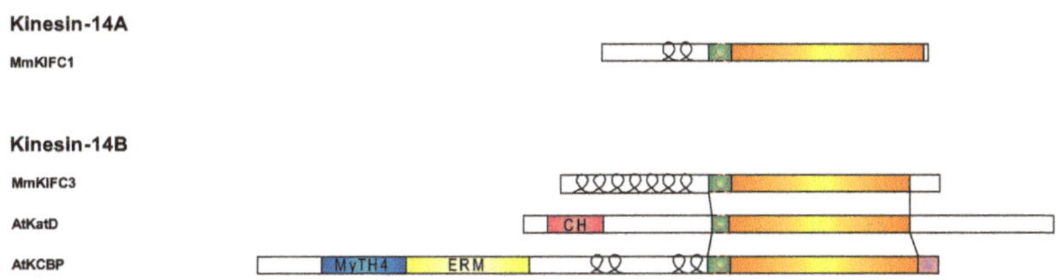

Figure 4.5 Structure of the kinesin-14 family. As in Fig. 4.2.

The kinesin-14A subfamily (previously been known as the C-1,[207,241] KIN C-Mitotic,[242] and C-I[244]) are characterised by a family-specific neck helix before the catalytic core at the C-terminus.[221] Members have been reported to have functions in nuclear fusion in yeast[292] and multiple functions in higher organisms,[291] with members spread across all four kingdoms. Interestingly, the Pkl1 kinesin-14A member from fission yeast appears to be a slow (33 ± 9 nm/s) minus end-directed microtubule motor.[293] Both cyro-electron microscopy (cyro-EM) imaging[294] and FRET studies[295] have shown a rotation of the coiled-coil stalk during ADP binding initiates the steps along the microtubule. Optical trap measurements of power strokes of the *Drosophila* kinesin-14 NCD have shown that, unexpectedly, 30% of apparent power strokes were plus-directed, while the majority were as expected and minus-end directed,[296] which indicates that directionality is a complex process.

The kinesin-14B subfamily has roles in organelle transport and has a family specific helical neck region located adjacent to the N-terminal end of the catalytic core. This subfamily is further broken down into three subgroups: the KIFC2 subgroup, the KatD subgroup, and the KCBP (kinesin-like calmodulin binding proteins) subgroup.[221]

The KIFC2 subgroup contains KIFs found only in animal cells, whereas the KatD subgroup comprises multiple related members from plants and members have been found to have a calponin-homology (CH) domain[221] and have been hypothesised to function in endocytosis: retrieving vesicles that have undergone endocytosis in conjunction with the actin cytoskeleton and then transporting vesicles along microtubules in a retrograde direction as has been suggested for KIFC2 in animal cells.[297]

The KCBP subgroup members are closely conserved in plants and have a relatively loosely conserved member in sea urchins. It has been proposed that the KCBP may have evolved to compensate for the lack of cytoplasmic dyneins in plants,[298–300] as both members of this subgroup are retrograde motors.

4.1.3.15 Orphans

There are several KIFs that have not been assigned families due their distinct sequences that distinguish them apart from other KIFs. These members are from nematodes, protochordate, and a slime mould, in addition to unique KIFs from Tetrahymena, Giardia, and Leishmania.[221]

The abundance of KIFs from multiple species has allowed the characterisation of KIFs with a number of specific functions. Most can translocate along microtubules, and members or each family are capable of transporting different cargoes. With the availability of such a wide number of KIFs, with varying functions, their use in synthetic biology derived systems seems certain.

4.1.4 Structure and Mechanism

The crystal structure of the conventional kinesin head domain from rat brain, from the kinesin-1 family, was shown to comprise a globular motor domain and a neck linker[301] (Fig. 4.6).

The N-terminal globular domains were identified as motor regions responsible for microtubule and nucleotide binding,[208,303] with a specific neck linker identified as being responsible for mechanical transfer of energy upon ATP hydrolysis.[219] The specific docking and undocking of the neck region give rise to the power stroke.[204] In addition, experiments showed that the neck linker conformation was responsible for determining the directionality of the motor upon the microtubule.[216,219,304]

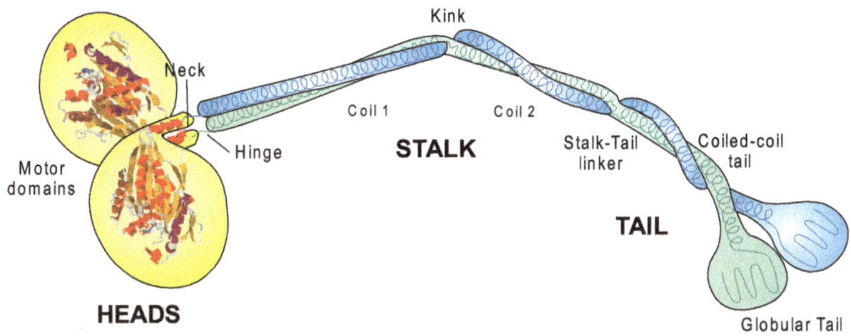

Figure 4.6 Domain architecture of the kinesin-1 heavy chain homodimer. The crystal structure of kinesin-1 from rat brain yields the N-terminal motor domain and neck linker,[301] with the remainder of the structure determined from structural prediction analyses coordinated with EM studies. Adapted from Woehlke and Schliwa.[302]

Following the neck linker lies an extended coiled-coil region responsible for dimerisation of the kinesin molecule,[305] which leads to a globular tail domain responsible for binding of the kinesin light chains.[242] The kinesin light chains, in turn, are responsible for regulation of the cargo-binging affinity.[208]

In the absence of cargoes, conventional kinesin undergoes autoinhibition by folding into an inactive state by interaction of the tail region with the motor domain. The autoinhibition is mediated via a conserved sequence motif (QIAKPIRP) in the tail that directly interacts with the Switch I helix in the motor domain, which prevents release of ADP from the ATP binding pocket.[306,307] Activation of the autoinhibitory state of kinesin-1 proteins into an active form requires two proteins — FEZ1 and JIP1 — which bind to the two inhibitory regions of the complex within the globular tail and the light chain subunit.[308] In addition to this, phosphorylation and also the nature of the cargo bound appears to have an effect on both the activity and also nature of the translocation (diffusive or processive).[309]

Single-molecule studies have elucidated that the kinesin-1 dimer takes 8.3 nm steps per ATP hydrolysis cycle, equal to the distance between tubulins,[310–313] and can complete ~100 ATP cycles (highly processive) with translocation reaching speeds of 800 nm per second and generating forces up to ~6pN.[311,314]

Two models for the translocation of kinesin-1 on microtubules have been proposed, the hand-over-hand model[204,315] and the inchworm model,[316] although the majority of the available data favours the hand-over-hand mechanism of movement.

The inchworm mechanism was based on the results of a microtubule gliding assay[316] and proposed that there was only one active motor head during each step, with one ATP hydrolysis giving rise to the movement of both heads forward one step without rotation of the stalk during the step.

Contrastingly, the hand-over-hand mechanism proposes a scheme whereby kinesin walks with one head remaining bound (stationary), whilst the other moves forward on the microtubule. This would result in each head domain moving 16.6 nm (and, therefore, the cargo 8.3 nm) per ATP hydrolysis cycle.[317] Based on a number of experiments,[318–320] it has clearly been shown that a hand-over-hand mechanism of translocation is more likely, with each step coordinated to the hydrolysis of one ATP molecule by each motor.[321]

This mechanism also raised the question of the coordination of the steps, whether they were symmetric or asymmetric. Kinesin-1 has been immobilised on a microtubule and imaged using atomic force microscopy at a resolution of 1 nm, which shows that a single kinesin dimer binds to the same protofilament of the microtubule (and does not straddle two protofilaments), and at submicromolar concentrations of ATP, both heads are frequently in contact with the microtubule.[322]

If the steps were symmetric and the heads passed around the stalk on the same side, each step would result in the rotation of the stalk and, therefore, cargo by 180° effectively winding the stalk until the torsional barrier prevented further steps.[323] If the steps were symmetric and the heads passed the stalk from alternate side, the stalk would be rotated 180° during each step resulting in a net 0° rotation allowing for the observed processive movement.[317]

The asymmetric hand-over-hand model proposed involves little or no rotation of the stalk and, therefore, cargo during the steps. This mechanism proposes that the neck linker winds around the stalk during the first step and then unwinds during the second step, with two different conformations of kinesin. The resulting translocation would result in kinesin having a limp, and recent FRET experiments have shown this to be the most likely mechanism.[324]

4.2 Myosin Superfamily

4.2.1 Function

Myosins are important in a great number of cellular processes, including muscle contraction, vision, hearing, cell motility, and host cell invasion of apicomplexan parasites. In contrast to the kinesin and dynein superfamilies that translocate along microtubules, myosin motors have been shown to mainly translocate actin-based filaments, although recent research suggests that certain members of the myosin superfamily members may undertake an ATP-independent "scan" along microtubules for relevant kinesin cargoes.[325]

As with the other motor proteins reviewed so far, the myosin superfamily members all require ATP to translocate actin filaments, with constituent families each having slightly different properties and functions reflecting their role. The majority of the families also exhibit a common directionality towards the plus end of the substrate.[326]

4.2.2 Myosin Complex

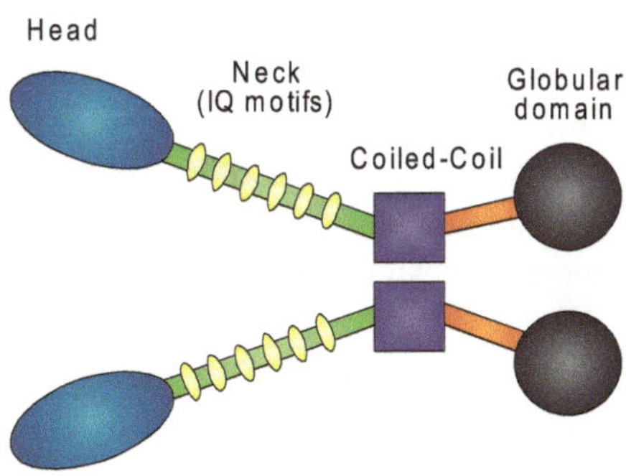

Figure 4.7 General structure of Myosin-V. Myosin-V consists of two heavy chains containing the myosin head and a neck with six IQ motifs each of which binds a total of four calmodulin molecules and one each of an additional 25 kDa and a 17 kDa light chain. Adapted from a figure in reference [330].

The myosin complex comprises the myosin heavy chain (either as a monomer or dimer), which often interacts with specific myosin light chains and a specific cargo. In general, the myosin heavy chain comprises an N-terminal, conserved, head domain, which is responsible for ATPase activity, actin filament binding, and motor activity. This is followed by a neck domain comprising a variable length helix dependent on the number of IQ (IQxxxRGxxxR) motifs (from none to eight), which are involved in interactions with light chains or calmodulin. Finally, there is a C-terminal, highly variable, tail region, which is generally responsible for specific cargo binding, cell localisation, and/or heavy domain dimerisation, which, therefore, determines the overall function of the specific myosin complex.[327–329]

4.2.3 Nomenclature

As with the other motor proteins discussed in this book, a wide variety of myosin genes, found throughout the eukaryotic phyla, were named according to different schemes as they were discovered. This has led to a somewhat confusing nomenclature that can make it difficult to compare the functions of myosin proteins within and between organisms.

A standardised nomenclature is yet to be determined across the myosin superfamily; however, a standardised nomenclature for the myosin-I class has been suggested whereby the abbreviation for the Linnaean binomial precedes the gene or protein name, e.g., the rat myosin, myr1, should be named RnMyo1b.[331] Indeed this same nomenclature has also been suggested for use in the naming of the members of the recently designated myosin classes XXII, XXIII, and XXIV.[332]

4.2.4 Classification

Through various phylogenetic analyses[332,333] and comparisons of the primary structure of the myosin heavy chain motor domain from different species, a superfamily of 26 different classes has been categorised.[332] The majority of experimental data regarding myosins has been obtained through work on the "conventional" class II myosin, which is found in muscle, with less known about the other "unconventional" classes.

4.2.4.1 Myosin-I

The first of the "unconventional" myosins to be discovered, *Acanthamoeba* myosin-I, revealed them to be single-headed myosins, unable to associate into bipolar filaments.[334,335] Subsequently based on phylogenetic analyses, the myosin-I family has been split into four subclasses:

4.2.4.1.1 Subclass-1 (or c)

Members of subclass-1 have been identified in *Acanthamoeba, Dictyostelium, Saccharomyces,* and *Emiricella* (*Aspergillus*) and vertebrates (chicken, mouse, rat, and human).[334,336] Primary sequence analysis shows a single motor domain, followed by a linker (neck) region comprising one or two IQ motifs required for calmodulin-like subunit binding. The remaining tail regions contain one to three Tail Homology Domains (THD1-3) depending upon the specific mysosin-I subclass-1 isoform. Long-tailed isoforms contain all three TH domains:

- The TH1 domain comprising a high number of basic residues that lies immediately after the linker region.
- The TH2 or GPA/GPQ domain comprising a high proportion of glycine, proline, and either alanine or glutamine capable of ATP-independent actin binding.
- The TH3 domain, homologous to the SH3 domain.[337]

In contrast, the short-tailed isoforms only contain the TH1 domain. Myosin-I subclass-1 myosins have been implicated in cell crawling, chemotaxis, phagocytosis, pseudopod extension, and contractile vacuole function[338] and experiments on these three subclasses and one myosin from *Dictyostelium* have shown them to have both shared and separate functions between them.[339]

4.2.4.1.2 Subclass-2 (or a)

Members of this subclass have been identified from *Dictyostelium* and vertebrates, where they have been found to be mainly expressed in gut, brain, lung, kidney, and liver tissues.[336] They differ from subclass-1 members in the number of IQ motifs (3–6) and only have the TH1 domain in their tail region; the basic-residue-rich sequence is thought to be important in localisation to microvilli where they act as cross-links between membrane and actin filaments.[340]

4.2.4.1.3 Subclass-3 (or b)

Originally discovered in vertebrates and later in *Drosophila* and vertebrates, the subclass-3 members contain a neck region with three IQ motifs and a TH1-type tail region containing multiple basic residues, which have been shown to bind phospholipids.[341] The members of the subfamily are expressed in a broad range of tissues, with the highest levels found in spleen, heart, lung, adrenal gland, oesophagus, and stomach. Additionally, they have been found associated with actin-rich structures in the plasma membrane (ruffles and protrusions) in cultured neurones,[335] and frog. Myosin-Ib has been shown to be localised to frog sacculus sensory hair cells where it is thought to modulate the sensitivity of the hair cell bundle to mechanical stimulus.[342]

4.2.4.1.4 Subclass-4 (or g)

Typified by the rat, myr4, and *Drosophila* myosin, DmIA, the subclass-4 members have two IQ motifs in their neck regions and their tail regions contain one TH1 domain, which is rich in basic and hydrophobic residues.[343] The myr4 protein is expressed in a wide range of tissues with the highest levels found in the adult brain, whereas the DmIA is expressed exclusively in the gut epithelia.[344]

The myosin-I class is the largest single class of myosins identified, but despite this, the majority of their functions remain to be elucidated. Evidence shows that the members of this class have an important role in the endocytic pathway and have been observed to be localised to the recycling endosome or lysosome.[345]

In vivo experiments have shown that regulation of myosins is through heavy chain phosphorylation, calcium binding to calmodulin in the neck region, or possibly by tail domain receptor targeting.[335]

4.2.4.2 Myosin-II or "conventional" myosins

Myosin-II genes have been identified in all eukaryotes so far examined, except plants. The myosin-II class can be broken down into four subclasses based on comparative analysis of the motor and tail region sequence:

- Skeletal/cardiac muscle (sarcomeric) myosins.
- Vertebrate smooth muscle/non-muscle myosins.
- *Dictyostelium/Acanthamoeba*-type myosins.
- Yeast-type myosins.

The majority of work to date has focused on the vertebrate striated muscle proteins,[346] and much less is known about the other groups of myosin-IIs.

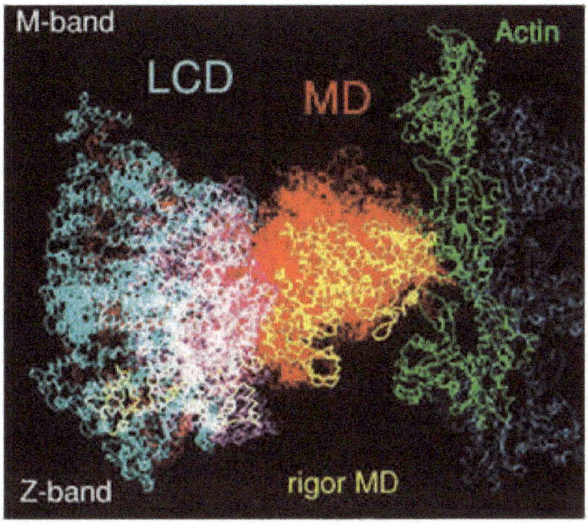

Figure 4.8 Rotation of the lever arm in myosin. An overlay of several images of the relative position of the myosin lever arm, obtained from crystallography data, which shows the rotation of the arm relative to the myosin main body. Reproduced from http://www.mrc-lmb.cam.ac.uk/myosin/Structures/Tomography.html with permission of Prof. Ken Taylor.

Structurally, myosin-II complexes comprise two heavy chains between 170 kDa and 240 kDa, two pairs of light chains (16–23 kDa), the regulatory (RLCs), and essential light chains (ELCs) that stabilise a length of helix in the neck region responsible for forming a "lever arm".[327]

The proposed model of myosin movement involves ATP-induced changes in the core motor domain (whilst attached to actin) that are propagated along the myosin heavy chain to the "converter" and "lever" domains resulting in them rotating relative to the catalytic domain (Fig. 4.8). This power stroke mechanism results in movement of the myosin against the actin filament.[329,347]

Regulation of myosin-II interaction with actin filaments has been shown to be mediated by four distinct mechanisms:

Phosphorylation of the RLCs in vertebrate smooth muscle and non-muscle myosin-II members by a calcium/calmodulin-dependent light chain kinase can switch on the myosin–actin interaction and also regulate assembly of myosin filaments.[348,349]

Vertebrate striated and cardiac muscle myosin-IIs are phosphorylated by a similar kinase; however, regulation is primarily controlled through the toponin and tropomyosin complexes on the actin filaments.[350]

Calcium-binding to the ELCs regulates myosin–actin binding in molluscan muscle myosin-IIs.[351]

Acanthamoeba and *Dictyostelium* myosin-IIs are phosphorylated at distinct sites of their heavy chain tails, which causes inhibition of actin binding, and further evidence suggests that RLC phosphorylation has an additional regulatory role.[348]

Although myosin-IIs are primarily known for their role in contraction and force production in skeletal, cardiac, and smooth muscles, studies on *Dictyostelium*, *Caenorhabditis elegans*, *Drosophila*, and yeast have indicated that myosin-II members are also essential for cytokinesis, cell motility, cell polarity and chemotaxis, cell architecture, and development in non-muscle cells.[352]

4.2.4.3 Myosin-III

First discovered in *Drosophila*, and named NinaC, myosin-III was determined to be a gene involved in a phototransduction defect.[353] Further research has led to the discovery of the myosin-III gene in horseshoe crabs, fish, and humans.[354]

NinaC has been shown to be monomeric; in contrast to all other myosins, it has an accessory domain in the form of an N-terminal kinase domain. Two splice variants exist in *Drosophila*: p174 has two IQ light chain binding motifs and a 420 amino acid tail that localises it to dense arrays of microvilli that comprise the rhabdomere of the photoreceptor of the eye. Mutagenesis of the gene has identified regions important for localisation, termination of phototransduction, and rhabdomere maintenance.[355–357] In comparison, the other variant, p132, has a single IQ light chain binding motif and a shorter 54 amino acid tail, and studies with this gene imply that this variant is not essential for phototransduction.[355]

4.2.4.4 Myosin-IV

Only one member of the myosin-IV class has so far been identified and that is from *Acanthamoeba*. Structure predicted from the primary amino acid sequence suggests a single-motor domain, with one IQ motif, and a Myosin Tail Homology (MyTH4) domain (implicated in microtubule binding),[358] which is homologous to that found in members of the myosin-VII and -XV classes.

4.2.4.5 Myosin-V

Present in most eukaryotes, except plants, the myosin-V class[359,360] can be organised into three distinct subclasses in vertebrates, myosin Va, Vb, and Vc[338,361] and two subclasses in yeast, Myo2p and Myo4p.[362] All subclasses are differentially expressed in organisms; however, in vertebrates it is not clear whether members from the different subclasses have distinct or overlapping functions *in vivo*.[338,361]

Myosin-V is a dimeric protein comprising a highly conserved motor domain followed by a region containing six IQ motifs required for binding light chains and calmodulin.[360,363] The downstream tail domain has been shown to be important for cellular localisation and cargo binding and in vertebrates has a specific α-helical coiled-coil region containing a PEST site[364] (calpain protease sensitive site) followed by a C-terminal globular region[365] (as shown in Fig. 4.7).

Experiments to elucidate the motor activity of myosin-V have shown the complex to be highly processive with what has been described as a high duty cycle (the time the myosin remains in contact with the actin fibre during movement), with myosin-V exhibiting a 51% duty cycle[366] compared to smooth muscle myosin-II, which was estimated to be 3%.[367] Step sizes for myosin-V have been measured at around 36 nm,[368,369] which is consistent with electron microscopy data that has allowed observation of the two motor domain (heads) binding at similar intervals.[370] Similar snapshots of the binding of myosin-V to an actin filament have allowed a model for movement to be developed.[371] The mechanism proposed starts with both heads of the myosin bound to the actin filament, binding of ATP causes detachment of the relevant myosin head, which relieves strain in the bound molecule bringing the detached head to the next actin site. The head attaches to the actin site, followed by loss of the phosphate by actin activation, which tightens the binding. In contrast to this data, however, optical trap experiments have reported step sizes of 20 nm, with occasional 37 nm step sizes.[366]

Functionally, the myosin-V family has been implicated in melanosome transport in melanocytes through mutagenesis experiments in mice, where myosin-V deficient mice exhibited loss of coat colour and neurological disorders.[372,373] Further experiments using time lapse video microscopy have highlighted a coordination between microtubule motors (kinesins) and myosin-V, with microtubule-based motors transporting melanosomes from the cell body to the malanocyte dendrites and myosin-V mediating their movement at the cell periphery on actin filaments.[338]

A similar type of coordination has also been observed in fish and frog melanophores, which undergo fast, coordinated aggregation and dispersion of pigment granules along microtubules. Using a similar technique to that used by Wu *et al.*, Rodionov *et al.*[374] showed that alongside the faster microtubule-based dispersion, short range actin-dependent movement of granules occurred between microtubules.

In yeast, the two myosin-V subclasses have differing roles. Myo4p has been shown to be required for the transport of maternal mRNA (ASH1) to the daughter cell to maintain mating type[375]; whereas, Myo2p has been implicated in vacuolar inheritance and vesicle trafficking in exocytosis.[362] Mutation of Myo2p results in defects in secretion, which has been interpreted as showing that myosin may be responsible for movement of secretory vesicles. Additionally, there is also evidence of a coordination between myosin-V and kinesin in yeast with the temperature-sensitive mutant myo2p66 suppressed by SMY1, a kinesin-like protein.[362]

4.2.4.6 Myosin-VI

First discovered in *Drosophila,* and later in most animal tissues, myosin-VI in most organisms (except in *Dictyostelium* and yeast) is a 140 kDa protein that dimerises via a single IQ motif in the neck region, with a tail region comprising a coiled-coil region followed by a globular domain.[376]

In contrast to other myosins, myosin-VI travels towards the minus end of actin filaments, which has been shown to be due to a specific converter region between the motor domain and neck region, termed the "reverse gear"[377,378] and function both as a transporter and as a cytoskeletal anchor *in vivo*.[379]

Drosophila myosin-VI has been shown to have a role in movement of cytoplasmic particles, formation of pseudocleavage furrows in embryos, spermatogenesis and also of egg chamber and imaginal disc morphogenesis.[380] Myosin-VI proteins also have the potential for coordination with microtubule-binding proteins, and in *Drosophila*, myosin-VI has been shown to co-immunoprecipitate with D-CLIP-190, an orthologue of human CLIP-170, which is a known microtubule-binding protein.[381]

Mammalian myosin-VI can be phosphorylated by a PAK kinase, which is under the control of Cdc-42 and Rac (small GTPases that mediate extracellular signals), which could suggest a link with intracellular signalling pathways.[382] In polarised cells, myosin-VI has been shown to be associated with the apical domain, whilst in sensory hair cells of the inner ear, myosin-VI is localised to the cuticular plate and pericuticular necklace.[383] Mutation of the myosin-VI gene in mouse (Snell's Waltzer) results in fusion of the stereocilia and deafness.[384]

4.2.4.7 Myosin-VII

The mysosin-VII isoform has been identified in *Drosophila*, *C. elegans*, pig, mouse, and human.[352] Mutations within the gene lead to loss of hearing in both mice (shaker 1) and humans (Usher Syndrome type 1B, which has associated eye function impairment, DFNB2, and DFNA11).

Based on sequence analysis, the protein is predicted to be dimeric, with five IQ motifs and a tail comprising a short coiled-coil region, two myosin-tail homology (MyTH4) domains, two talin-binding (FERM) domains, and an SH3 domain.

Experiments using total internal reflection fluorescence (TIRF) microscopy to study the *Drosophila* myosin-VII showed it to be a processive motor with a step size of 30 nm, evidence that supports a hand-over-hand mechanism of motor movement as has been suggested for myosin-V.[385,386]

In the inner ear sensory cells, myosin-VII is thought to act as a conveyer of several Usher syndrome proteins that are involved in the differentiation of the hair bundle. In the retina, myosin-VII transports melanosomes and phagosomes in pigment epithelium cells, and opsins in photoreceptor cells. In *Dictyostelium*, myosin-VII has been observed to be required for phagocytosis[387] and cell adhesion.[388]

4.2.4.8 Myosin-VIII

The first of the plant myosins to be discovered was the ATM1 myosin from *Arabidopsis thaliana*.[389] Subsequently, three further myosin-

VIIIs were discovered in *Arapidopsis* (ATM2, myosin-VIIIA, and myosin-VIIIB) and the myosin-VIII class now has seven members. The protein comprises an N-terminal PEST domain (a site of calpain cleavage) prior to the conserved motor domain, followed by four IQ motifs and a short region of predicted α-helical coiled-coil, which could be required for dimerisation.[390] At the C-terminus, clusters of basic residues are thought to be involved in phospholipid binding.[389]

Myosin-VIII has been determined, by immunostaining, to be mainly localised to the plasma membrane at cytokinetic cell plates and plasmodesmata.[391] More recent experiments using transgenic *Arabidopsis* have shown that myosin-VIII functions differently in different root cells and could potentially be involved in steps of endocytosis, vesicle trafficking inhibitor brefeldin (BFA)-sensitive and insensitive pathways, ER tethering as well as plasmodesmatal activity.[392]

4.2.4.9 Myosin-IX

Identified in vertebrates and also the nematode *C. elegans*, the myosin-IX members are widely expressed in many cell types and exhibit a similar directionality to that of the myosin-VI class, travelling towards the minus ends of actin filaments. Myosin-IX appears to be more widely expressed in cells in comparison to myosin-VI and, therefore, it has been assumed that the majority of minus-end myosin migration is due to this class. Localisation of the protein tends to be to the cytoplasm although it has also been found partly associated with vesicles, the plasma membrane, and microfilaments.[393]

Invertebrates have only a single instance of the gene; whereas, vertebrates have two forms. The human myosin-IXA are thought to be the cause of Bardet–Biedl syndrome, an autosomal recessive disorder causing obesity, short stature, retinitis pigmentosa, and mental retardation.[394]

The structure of the monomeric protein comprises an N-terminal extension of unknown function followed by the motor domain, four IQ motifs, followed by a zinc-binding motif and a C-terminal Rho/Rac GTPase activating domain within the tail of the molecule.[393,395,396] This GTPase domain implicates the protein as having a role in the activation of the Rho pathway leading to myosin-II activation.

Although the protein can be considered single headed (monomeric), experimental data suggest that myosin-IXb is highly processive[397,398] and remains bound to actin filaments even in the presence of ATP, which could potentially imply the presence of a secondary actin-binding site in the molecule.[395]

4.2.4.10 Myosin-X

Myo10, the founding member of the myosin-X class, was discovered in frog inner ear[399] and since then has been found in bovine tissue. Myosin-X is not present in either *C. elegans* or *Drosophila*, implying it may be vertebrate specific. Not much is known about the oligomerisation state of myosin-X; however, the protein has a head domain comprising the motor, a neck region with three IQ motifs, and tail region comprising initially a coiled-coil region (possible for dimerisation), three PEST regions, three pleckstrin homology (PH) domains (responsible for inositol phospholipid binding), an MyTH4 domain (involved in microtubule binding[358]) and a C-terminal talin-like FERM domain (required for integrin binding).

Myo10 is only expressed in low levels in the cell and has been found to be concentrated at the tips of filopodia.[400] Myo10-rich regions of actin filaments have been observed to travel forward and backward along filopods, although the specific function of Myo10 at the tips of filopodia remains to be determined. The forward movement has been proposed to be due to the motor activity of Myo10, whereas the backward movement is thought to be due to tight binding of Myo10 to actin filaments that are undergoing retrograde flow.[401]

4.2.4.11 Myosin-XI

Unique to algal and higher plant cells, the myosin-XI class comprises more than 18 members. The structure of the protein bears similarities to that of the class V myosins,[402] with an N-terminal motor domain, followed by the neck region that contains five or six IQ motifs, and a long C-terminal tail region[403] comprising predicted coiled-coil regions and dilute domains (implicated in dimerisation and cargo binding).

The myosin-XI members exhibit directionality towards the positive end of actin filaments and appear to be highly processive, with translocation speeds of up to 60 mm per second and a step size of 35 nm.[404–406] Evidence suggests that they are responsible for the rapid organelle transport observed in cytoplasmic streaming,[407] as well as being essential for tip growth,[408] and Golgi, peroxisome, and mitochondrial movement.[409]

4.2.4.12 Myosin-XII

Considered an "oddity" due to the low level of sequence conservation, there is one member of the myosin-XII class identified from *C. elegans.*[326] Based on phylogenetic analysis of the similar tail regions, it has been proposed that they share a common ancestry with the class XV myosins.[332] The gene encodes a protein of ~300 kDa, with an N-terminal extension of 150 amino acids followed by a weakly conserved head domain, followed by a neck region that has two IQ motifs, and a tail region that has two MyTH4 domains.

4.2.4.13 Myosin-XIII

As yet there are only two members of the myosin-XIII class, both discovered in *Acetabularia cliffonii*, a green algae.[410] Structurally myosin-XIII members follow a similar domain structure to other myosins, with a head domain, followed by a neck region containing up to seven IQ motifs (Alcmyo2), and a short tail region (Alcmyo1). Confocal scanning laser microscopy shows these myosins to be localised to organelles, the budding region of lateral whorls, and the cell apex, suggesting their involvement in organelle transport and tip growth.[410]

4.2.4.14 Myosin-XIV

Originally populated by specific myosins from the parasites *Toxoplasma gondii*[411] and *Plasmodium falciparum*,[412] the class has now been extended to include myosins from the ciliate *Tetrahymena thermophile*,[332] and the class has been split down into four subclasses (a–d), dependent on their predicted domain structure.

Myosin-IVa members are apparently the simplest myosins discovered to date, having an N-terminal head domain, followed by a maximum of one IQ motifs in the neck region, and no tail region. Myosin-IVb members have a similar structure to the IVa subclass and are differentiated based on sequence analysis. The myosin-IVc subclass differs significantly with a head domain followed by six to eight IQ motifs and a tail region with a domain similar to the α-tubulin suppressor (ATS1). Finally, the myosin-IVd subclass comprises 12 ciliate myosins from *Tetrahymena thermophila.* These have the familiar head domain, followed by up to three IQ motifs, and a tail region with either an ATS1 or MyTH4 domain.[332]

4.2.4.15 Myosin-XV

Myosin-XV is single headed and is the largest myosin so far identified at ~395 kDa. It comprises a long N-terminal extension of ~1,200 amino acids followed by the head domain, two IQ motifs, and a tail domain closely related to that of the myosin-XII class members.[413]

The myosin-XV class was originally populated with genes from humans and mice, which were determined to be responsible for hereditary deafness DFNB3 and *shaker* 2, respectively,[413–415] where the protein appears to be localised predominantly to the cell body and the stereocilia of hair cells in the inner ear and pituitary gland. Subsequently, a myosin-XV homolog has been identified in *Drosophila*,[416] where it is required for proper dorsal closure, a morphogenetic event that occurs late in *Drosophila* embryogenesis.[417] Specifically, it is required for trafficking sensory, cytoskeletal, and adhesion cargos within leading edge cells and their related filopodial protrusions.[418]

4.2.4.16 Myosin-XVI

The founding members of this class, two distinct isoforms Myo16a and Myo16b, were identified in rat[419] and have subsequently been identified in a number of other vertebrates.[333,420] The predominant isoform, Myo16b, has an N-terminal ankyrin repeat domain (this is thought to mediate an association with protein phosphatase 1 catalytic subunits 1α and 1γ), followed by the motor domain, a single IQ motif, and a long C-terminal tail region.[421]

Myo16b appears to be localised primarily to neuronal cells throughout the developing and adult central nervous system, with lower expression levels detected in some peripheral tissues, including lung, kidney, and several endocrine tissues.[419] More recent studies have shown that Myo16b is targeted to the nucleus of cells transitioning through interphase, via a nuclear localisation domain in the C-terminal tail of the protein.[421]

4.2.4.17 Myosin-XVII

The class XVII myosins have two members, both chitin synthase myosins (Pg csm1 and En csmA), assigned based on phylogenetic sequence analysis.[422] These proteins comprise an N-terminal myosin head coupled with a C-terminal chitin synthase domain, which is essential for maintaining the integrity of the cell wall in filamentous fungi. Filamentous fungi utilise an actin cytoskeleton, and it is proposed that CsmA transmits cell polarity to the localised synthesis of the cell wall at the hyphal tips and septation sites via its association with the actin cytoskeleton.[423,424]

4.2.4.18 Myosin-XVIII

Closely related to, if not part of, the myosin-II class,[332] myosin -XVIII is split into two subclasses, A and B. First isolated from bone marrow stromal cells of mice and named MysPDZ,[425] myosin-XVIIIA has subsequently been identified in humans and flies.[416,426,427] Myosin-XVIIIA is predicted to be a two-headed motor and has two isoforms (designated α and β in humans and mice), which differ primarily on the presence of their N-terminal region ahead of the motor domain. The α isoform has an N-terminal region containing a PDZ domain[428] (known to act as a scaffold for holding together signalling protein complexes); whereas, the β isoform has no N-terminal region prior to the motor domain.[425,426] The overall structure of both isoforms comprises the isoform-dependent N-terminal region, where present, followed by the motor domain, a neck region of incorporating one IQ motif and a tail domain that has a coiled-coil region followed by a globular domain. Experiments with the N-terminal PDZ domain of the α isoform have shown it to have an ATP-independent actin-binding domain ahead of the PDZ domain, which has been proposed to be important in the higher-order organisation of the actin cytoskeleton.[429]

Myosin-XVIIIB has been identified in humans as well as other vertebrates and has a similar core structure to that of myosin-XVIIIAs. It differs at the N-terminal and C-terminal regions that, as yet, have no known function and this subclass appears to be unique to vertebrates. Research has shown that the C-terminus has nuclear localisation signals, and myosin-VXIIIB has been predicted to have roles in cellular trafficking as well as possible myosin gene regulation within the myonucleus.[430]

4.2.4.19 Myosin-XIX

Only fairly recently recognised as a class, based on a phylogenetic analysis,[332] the myosin-XIX class comprises specific myosins from chordates, including human, ascidian *Ciona intestinalis*, fish, chicken, and dog. The founding member Myo19 from humans is expressed in multiple tissues and cell lines and is localised to the mitochondria, implicating a role in mitochondrial transport and dynamics. Structurally the protein appears to comprise an N-terminal motor domain, followed by a neck region containing three IQ motifs, and a short C-terminal tail responsible for mitochondrial localisation.[431]

4.2.4.20 Myosin-XX

Another more recently assigned class, the myosin-XX class encompasses sequences from *D. melanogaster* (including the founding member Myo29D), *Drosophila pseudoobscura*, and the malaria parasite vector *Anopheles gambiae*. This class appear to be insect specific, and the members appear to have one conserved IQ motif in the neck region of the protein.[332]

4.2.4.21 Myosin-XXI

Recently identified as a class specific to parasitic kinetoplastids, the myosin-XXI class encompasses sequences from *Trypanosoma cruzi*, *Trypanosoma brucei*, and *Leishmania major*. These genes code for proteins that have quite different associated functions to other unconventional myosins and have been predicted to encode FYVE[432] (phosphoinositide-binding), WW-like[433] (mediate protein–protein interactions through proline-rich motifs and are functionally similar to SH3-domains), or UBA[434] (ubiquitin-association)-like domains in their tails.[332]

4.2.4.22 Myosin-XXII

Also recently classified by phylogenetic analysis, the myosin-XXII class comprises previously unclassified myosins from the Chromalveolate- encompassing apicomplexan MyoF sequences alongside myosins from the diatom alga *Thalassiosira pseudonana*. The apicomplexan class XXII myosins are predicted to exhibit an extended neck domain with multiple IQ motifs (similar to the class V and related class VIII, XI, and XIII[435]). Additionally for those sequences with a long tail, between four and six WD40 repeats[436] implicated in processes from signal transduction and transcription regulation to cell cycle control and apoptosis) have been predicted towards the C-terminus.[332]

4.2.4.23 Myosin-XXIII

The myosin-XXIII class comprises two MyoG genes from coccidians *T. gondii* and *Eimeria tenella*. Proteins have been predicted to contain between one and two IQ motifs as well as an MyTH4 domain in their tails.[332]

4.2.4.24 Myosin-XXIV

The founding members of the myosin-XXIV class are TgMyoI from *T. gondii* and a homolog from *Cryptosporidium parvum*. As with the myosin-XXII class member, the myosin-XXIVs are predicted to contain one or two IQ motifs in their neck domain as well as an MyTH4 in the tail region of the peptide.[332]

4.2.5 Structure and Mechanism

Structures of myosin-II from chicken and scallop fast skeletal muscle and vertebrate smooth muscle myosins with various nucleotide analogues bound have yielded a number of conformations of myosin-II; these have given insight into the mechanism of the motor domain and the way in which movement and force are generated with respect to its substrate actin.[327–329]

Structures of the myosin head domain comprise the motor domain (containing the actin-binding site as well as the ATPase site) and the neck region formed from extended α-helix (responsible for binding of essential and regulatory light chains) (Fig. 4.9). It is the neck domain, acting as a lever arm, that is proposed to be essential for amplification of mechanical force and movement generated through the ATPase cycle of the motor domain.[327,437,438] Myosin makes contact with the actin fibre in the presence of ADP. Release of this ADP produces a conformational change in the myosin resulting in a 10 nm shift in the relative position for protein–protein contacts between myosin and actin — a translation of the two proteins relative to each other. ATP binding results in release of myosin from the actin fibre and hydrolysis of ATP to ADP, resulting in a reorientation of the myosin with respect to the actin, allowing rebinding of the myosin to the actin. Following this the whole cycle can repeat.

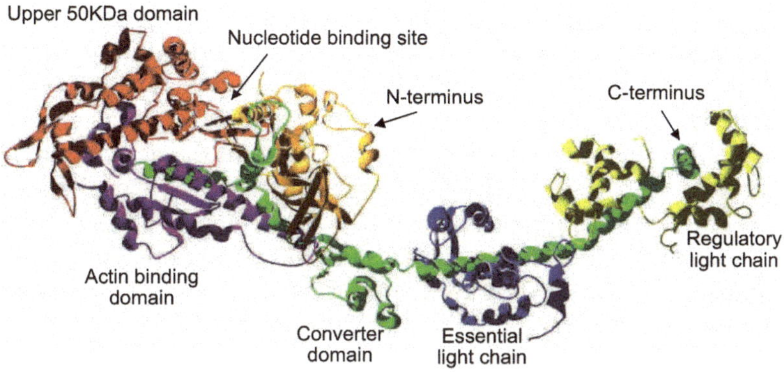

Figure 4.9 Ribbon representation of cryofixed myosin subfragment-1 from insect flight muscle. Ribbon representation of insect myosin subfragment-1 as depicted by Wu *et al.* (PDB entry: 2W4H), depicting the positions of the main upper 50 kDa domain (red) as well as position of the proposed nucleotide-binding site, the actin-binding domain (purple), converter domain, essential light chain (blue), and regulatory light chain (yellow).

Structures of a truncated *Dictyostelium* S1 fragment with ADP. BeF_3 and ADP.vanadate show the open (pre-power stroke) and closed (post-power stroke) conformations of the myosin heavy chain (Fig. 4.8). The difference in the structures lies in the 70° rotation of the converter domain relative to the motor, and it was these observations that confirmed a "swinging lever arm" model[347,439] to explain the myosin power stroke.[440]

Although these experiments have given us insights into the movement of the lever arm during myosin motor activity, due to the lack of an actomyosin complex structure, we still do not know the exact rearrangement of residues in the head domain upon actin binding and the effect this has on the power stroke during the cross-bridge cycle of conventional myosin activity. However, recent studies of a myosin-I from sea scallop suggest a hinge system, which may function in the "on" and "off" states modulating flexibility of the lever arms,[441] and novel time-resolved FRET studies have provided information on structural kinetics.[442]

Similar structural experiments to those conducted on myosin-II have been used to investigate the unconventional myosin-V, yielding similar results with an effective rotation of domains with respect to each other, dependent on whether they form crystals of the protein in the open or closed states. The structures also revealed the cooperativity between the actin and nucleotide-binding sites in relation to the binding of their relative substrate (actin or ATP).[443] Studies also highlighted the importance of a surface loop with net positive charge, Loop2, of the actin binding interface that appears to be involved in a weak electrostatic interaction with actin. Mutations of this loop region to either reduce or increase positive charge clearly showed that the processivity of the enzyme is significantly affected causing either shorter or longer runs, respectively, in relation to the wild type activity.[444]

Interestingly, myosin-V appears to have an inhibited state, with the molecule effectively folding in on itself with interactions between the tail region and the motor domain allowed by bending of the peptide between the lever and the rod regions.[445]

4.3 Dyneins

4.3.1 Function

Dyneins are only found in eukaryotes and have been lost from higher plants,[167] which appear to utilise retrograde translocating kinesins in their place.[446] In fact, evolutionary analysis has suggested that the loss of cytoplasmic dynein 1 (DYNC1) predates the divergence of the archaeplastida.[447]

There are two classes of dyneins that have distinctly different roles in the eukaryotic cell. Both classes are ATPases requiring ATP for activity and are members of the ATPase associated with various cellular activities (AAA^+) superfamily.[170] Despite their similarity in function to the KIFs, dyneins exhibit very different mechanisms and structures to kinesins. Dynein heavy chains form hexameric ring structures and have been suggested as having a winch-like mechanism of translocation mediated through a single monomer that is the site of ATPase activity.

There are two families of dynein complexes: the cytoplasmic and axonemal dyneins.[448] In contrast to the majority of the members of the KIF superfamily that translocate microtubules in an anterograde direction towards the plus end of microtubules, the cytoplasmic dyneins translocate microtubules in a retrograde direction towards the minus end of the microtubule.[449,450] Similar to KIFs, they conduct a range of functions, including transport of cargo (vesicles, organelles, chromosomes, protein complexes, and ribonucleoproteins), and they are also known to be involved in microtubule dynamics.[451,452]

The other family of dyneins is the axonemal dyneins that are mainly responsible for the sliding of microtubules using cilia and flagella to produce movement and these make up the majority of the dyneins so far identified.[451,453]

4.3.2 Dynein Complexes

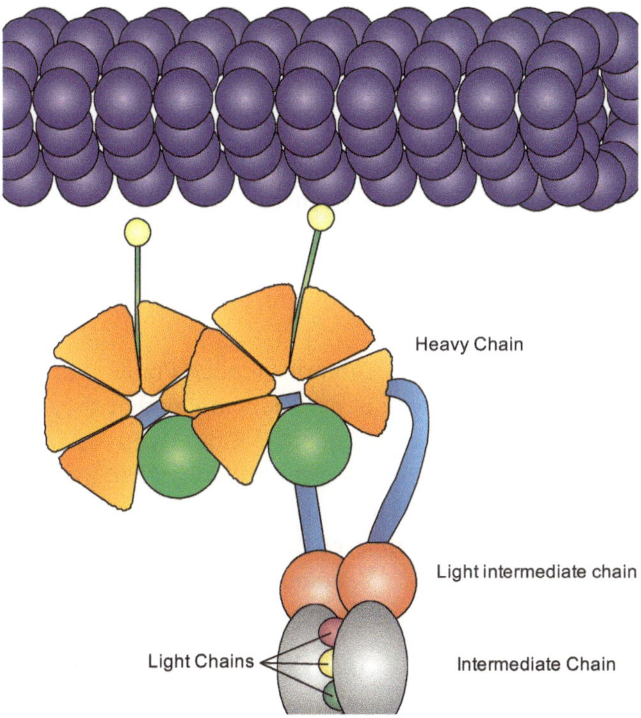

Figure 4.10 A dynein complex. Illustration of a generic cytoplasmic dynein complex encompassing the heavy chain, light intermediate, intermediate, and light chains. Redrawn from reference [453].

A dynein complex comprises 1, 2, or 3 dynein heavy chains, DYNCHs (alternatively known as DHCs, DYHs, or DYNs), which are large polypeptides (>500 kDa) and incorporate the motor domain of the dynein complex. The tail regions of these heavy chains can bind a number of smaller subunits: the light chains (8–30 kDa) or DYNCLs (alternatively known as LCs), the light intermediate chains (30–60 kDa) or DYNCLIs (alternatively known as LICs), and the intermediate chains (60–140 kDa) or DYNCIs (alternatively known as ICs).[454] Each of these lighter chains has a specific role in dynein complex activity and cargo binding (Fig. 4.10).[451]

4.3.3 Nomenclature

As with members of the kinesin superfamily, there has been some divergence in the naming of dynein subunits (leading to similar confusion) and this has recently led to a standardisation of the nomenclature for cytoplasmic dyneins based on mammalian cytoplasmic dynein subunits (Table 4.2), with the aim that the same nomenclature will be used for all dyneins, including axonemal dyneins.[455]

4.3.4 Classification

Classification of the dynein subunit families (Table 4.2) has been mainly based upon the type of dynein complex that they comprise and their cellular compartmentalisation (i.e., axonemal or cytoplasmic). The dynein heavy chains all appear to have a similar domain structure, comprising an N-terminal stem/tail region, linker region, and a six-membered AAA^{+} ring structure (responsible for the ATP hydrolysis)[456,457] amongst which is found the stalk region responsible for microtubule binding[458–460] (Fig. 4.11).

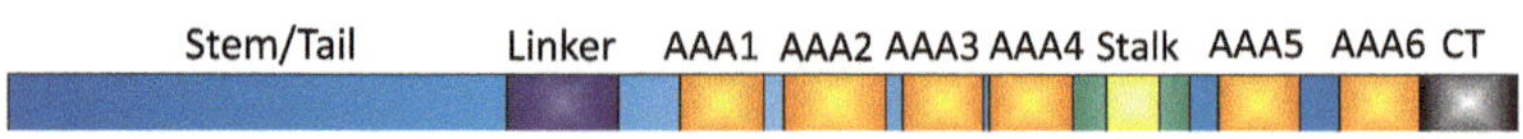

Figure 4.11 Domain structure of a dynein heavy chain. A bar representation of the domain structure of a dynein heavy chain, showing the stem/tail region (responsible for binding of HC, ICs, and LICs), linker region (implicated in force transduction), and the AAA units (1–6, responsible for ATP hydrolysis) as well as the stalk region (microtubule binding).

Table 4.2 Nomenclature for eukaryotic dynein complex subunits

Gene symbols				Revised protein symbols	Common protein names
Human		Mouse		Human	Human and
Revised	Former	Revised	Former	and mouse	mouse
Cytoplasmic dynein 1 complex					
DYNC1H1	DNCH1	Dynch1	Dnchc1	DYNC1H1	Cytoplasmic dynein 1 heavy chain, conventional cytoplasmic dynein heavy chain, DHC1a, dynein 1, dynein heavy chain 1, MAP1C
DYNC1I1	DNCI1	Dync1i1	Dncic1	DYNC1I1	Cytoplasmic dynein 1 intermediate chain 1, IC-1, IC74-1, IC70
DYNC1I2	DNCI2	Dync1i2	Dncic2	DYNC1I2	Cytoplasmic dynein 1 intermediate chain 2, IC-2, IC74-2
DYNC1LI1	DNCLI1	Dync1li1	Dnclic1	DYNC1LI1	Cytoplasmic dynein 1 light intermediate chain 1, LIC1
DYNC1LI2	DNCLI2	Dync1li2	Dnclic2	DYNC1LI2	Cytoplasmic dynein 1 light intermediate chain 2, LIC2
DYNLT1	TCTEL1	Dynlt1	Tctex1	DYNLT1	Dynein Tctex1 light chain 1, Tctex1
DYNLT3	TCTE1L	Dynlt3	Tcte1l	DYNLT3	Dynein Tctex1 light chain 3, rp3

(*Continued*)

Table 4.2 (*Continued*)

Gene symbols				Revised protein symbols	Common protein names
Human		Mouse		Human and mouse	Human and mouse
Revised	Former	Revised	Former		
DYNLRB1	DNCL2A	Dynlrb1	Dnclc2a	DYNLRB1	Dynein Roadblock light chain 1, Robl1
DYNLRB2	DNCL2B	Dynlrb2	Dnclc2b	DYNLRB2	Dynein Roadblock light chain 2, Robl2
DYNLL1	DNCL1	Dynll1	Dnclc1	DYNLL1	Dynein LC8 light chain 1
DYNLL2	DNCL2	Dynll2	Dlc2	DYNLL2	Dynein LC8 light chain 2
Cytoplasmic dynein 2 complex					
DYNC2H1	DNCH2	Dync2h1	Dnchc2	DYNC2H1	Cytoplasmic dynein 2 heavy chain, IFT1 dynein heavy chain, DHC1b, dynein heavy chain 2, DHC2
DYNC2LI1	n/a	Dync2li1	n/a	DYNCLI1	Cytoplasmic dynein 2 light intermediate chain 1, D2LIC, LIC3

Source: Reproduced from reference [461].

However, as mentioned previously, the dynein light, light intermediate, and intermediate chains cannot, as yet, be classified in such a simplistic manner due to the appearance of these subunits in multiple different dynein complexes and their classification is limited to their cellular compartmentalisation alone.

4.3.5 Axonemal Dyneins

4.3.5.1 Function

The axonemal dyneins, as their name suggests, are involved in movement of microtubules within axonemes, which are highly specialised arrays of microtubules found within flagella and cilia. These motile axonemes generally comprise 20 microtubules arrayed as nine fused pairs (the outer doublets) around two centrally located microtubules[462] (see insert of Fig. 4.12). The axonemal dyneins work by connecting the outer doublets and forcing them to slide against each other to cause a wave-like motion.[463]

The majority of work on axonemal dyneins has been conducted on *Chlamydomonas reinhardtii* (a biflagellate green alga), *Tetrahymena pyriformis* (ciliate protozoan), and also sea urchin sperm, in order to determine the exact way in which dyneins drive the flagella beat, whose exact mechanism remains elusive.[464] To date though, somewhat surprisingly, only limited work has been conducted on vertebrate forms.

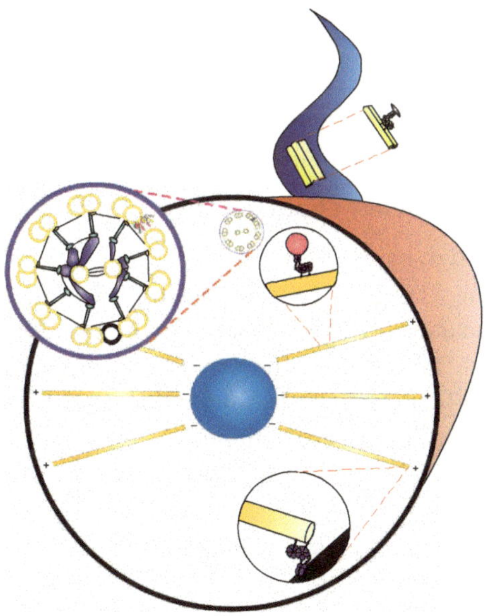

Figure 4.12 Positioning of the axonemes. Illustration of a flagellum, showing positioning of the axonemes and their associated microtubule arrangement. Redrawn from reference [453].

The axonemal dyneins have diverse roles in the movement of the axonemes, and this is reflected by the number of dynein genes associated with any given flagellum or cilia, with multiple biochemically defined forms found in any given flagellum or cilium.

Of those axonemal dyneins that have been characterised, all have been found to be stably associated with the complete microtubule of the outer doublet, aligned in rows along either the inner or outer edges. The dyneins found on the outside edge of the microtubules appear to have either three heads/motor domains (protozoans), or two heads/motor domains (metazoans); whereas those found arrayed in the inner surface of the microtubules only have one or two heads/motor domains.[454]

Within a flagellum, each dynein interacts with the neighbouring microtubule of the outer doublet and in studies on *C. reinhardtii*, seven subtypes of axonemal inner-arm dynein have been identified, with one forming a heterodimer and the rest found as single motor domains.[465]

4.3.6 Families

As yet, comparative genomic analysis of 24 diverse eukaryotes has led to the identification of seven axonemal dynein heavy chains (classified based on the dynein complex that they comprise) and also five axonemal dynein intermediate chains[447]:

Axonemal dynein heavy chain families

- Outer-arm dyneins
  - o OADα family (encompassing Tetrahymena OADα, metazoan OADα and Chlamydomonas OADγ)
  - o OADβ family (encompassing Tetrahymena OADβ and OADγ (orthologous to Chlamydomonas OADα)
- Two-headed inner-arm dyneins
  - o IAD-1α
  - o IAD-1β
- Singled-headed inner-arm dyneins
  - o IAD-3
  - o IAD-4
  - o IAD-5

Axonemal dynein intermediate chain families

- IC70
- IC78
- IC140
- IC138
- WDRD34

4.3.7 Structure and Mechanism

As previously mentioned, there are two classes of axonemal dyneins based on their positioning on the axoneme, which itself is composed of nine doublet microtubules in a ring surrounding a pair of single microtubules, named the central pair (CP) (Fig. 4.12). Each of the outer doublets of the axoneme is linked to its neighbour by strands of protein called the "nexin links"[463] and additionally has a series of projections called "radial spokes" that have been proposed to simply act as spacers that position the doublets in circle around the CP.[466] The CP is sheathed by a number of proteins that again form a series of projections that are presumed to interact with the radial spoke heads.[467] The spokes are anchored to the outer doublet in close proximity to the dynein regulatory complex (DRC),[468] which is itself in close proximity to the dynein inner-arms, and both of these have been shown to contain calcium-binding proteins that are known to be affected by the presence of free calcium and cause change in the beating pattern.[469] It is the heavily regulated sliding caused by the inner- and outer-arm dyneins of the microtubule doublet with respect to each other that creates the basic movement, which is converted to bending of the whole flagella and cilia by the mechanical restraints comprising the nexin links and basal anchor.[470–472] Several models have been proposed for the regulation and coordination of this complex movement, but as yet the exact answer remains elusive, although it seems from the available experimental evidence that control of dynein function and, therefore, microtubule sliding through the axoneme components located at the CP, spokes, and DRC is very likely.[464]

The best structural analysis of axonemal dynein complexes has been through cyro-EM studies of the interaction of the outer dynein arms with microtubules. Three-dimensional reconstruction of the images has allowed for insights into the function of this particular axonemal dynein. The study was conducted using reconstituted doublets made by cross-linking tubulin in the presence of axonemal dyneins from *C. reinhardtii*.[473] Results showed a conformational change between relaxed and rigor states (in the presence of ATP) of doublet pairs and allowed the authors to identify specific roles for each of the three heads of the outer-arm dyneins, with the α-head shown to be more flexible and assigned a regulatory role. The β-head was assigned the role of coordinating movement of the dynein with respect to the microtubule, and the γ-head was assigned as an anchor for the dynein to the microtubule. The stalk region, which binds to microtubules in a nucleotide-dependent manner, was present but could not be modelled due to its flexibility, and the dynein complex was observed to bind to the first microtubule primarily through a "base complex", which is thought to comprise two dynein intermediate chains.[474]

4.3.8 Cytoplasmic Dyneins

4.3.8.1 Function

Two forms of cytoplasmic dynein complex have been identified so far in human and mouse cells, named cytoplasmic dynein 1 and cytoplasmic dynein 2, with both complexes assembling as homodimers. However, their distributions through the cell, as well as their functions, differ greatly.

The cytoplasmic dynein 1 complexes are far more abundant and are found in all microtubule-containing cells, where they have roles in transporting components of the Golgi apparatus, lysosomes, and endosomes and have also been found to be responsible for retrograde axonal transport. In addition, the cytoplasmic dynein 1 complex has been found to associate with a diverse range of other complexes (both RNA and protein), which can be transported to, or can be maintained at, the microtubule minus ends. Experiments have shown that the cytoplasmic dynein complex is able to bind kinetochores where a role in microtubule capture has been indicated with the potential that the complex may remove checkpoint proteins to allow progression to anaphase. Complexes have also been found associated with the cell cortex during cell division and directed migration. These latter roles have led to the proposal that dyneins may have additional functions beyond that of a simple transport motor with the ability to create tension along or at the ends of microtubules from specific cell sites. Recombinant versions of these motors have recently been produced opening an opportunity for greater use of the proteins in bionanotechnology.[475]

In contrast to the wide dispersal of the cytoplasmic dynein 1 complex, the cytoplasmic dynein 2 complex is predominantly found localised within and around the base of cilia and flagella, where it is involved in retrograde intraflagellar transport, with a lesser role in the transport of proteinaceous rafts through modified ciliary structures such as the connecting cilium of photoreceptors.

Within the DYNCHs (Table 4.2), the ATPase domain is highly conserved, but outside of this motor domain, there is considerable variability, presumably reflecting the function of the particular heavy chain in specific cellular functions. The DYNCIs also show conservation of sequence between members of the family and exhibit some specificity towards specific dynein complexes, whereas the DYNCLCs exhibit the most variability and can be separated into several families that are not limited to defined dynein complexes, with some DYNCLCs found across several different dynein complexes.[447]

4.3.9 Families

Similarly to the axonemal dynein families, families have been identified and assigned based on comparative genomic analysis. As such two cytoplasmic dynein heavy chain families have been confirmed, one cytoplasmic dynein intermediate chain family, one cytoplasmic light intermediate chain family, and two cytoplasmic light chain families have been defined[447]:

Cytoplasmic dynein heavy chain families

- DYNC1H1
- DYNC1H2

Cytoplasmic dynein intermediate chain family

- DYNC1I1/2

Cytoplasmic dynein light intermediate chain family

- DYNC1LI1/2

Cytoplasmic dynein light chain families

- DYNLT1
- DYNLT2

4.3.10 Structure

The N-terminal domain of the dynein heavy chain comprises ~1,300 amino acid residues that form a relatively short tail region responsible for heavy chain dimerisation as well as accessory domain binding. This tail region is widely divergent between the heavy chains presumably reflecting the different purposes of the different dynein complexes which they comprise. The N-terminal tail region is followed by the six AAA modules from the core of the motor domain (numbered AAA1-6, respectively) amongst which is found the microtubule binding domain (MTBD).[476,477]

The first four AAA modules of the dynein heavy chain exhibit the highly conserved P-loop/Walker B motif (P1–P4, respectively) required for nucleotide binding. The first of these, P1, is absolutely conserved within dynein heavy chains and is the site for ATP hydrolysis,[478] required for movement, whereas the following three P-loops, P2–P4, are thought to be primarily involved in regulation of the dynein motor with variable and relatively low binding affinities for ATP[479]; although there have been some reports of AAA3 being an active ATPase.[480] The remaining two P-loops (P5 and P6) located terminally to the MTBD both appear to lack a nucleotide-binding site.

Of the P-loops, P3 appears to be the most important for dynamic regulation of the motor, with mutations in this region causing problems with release from microtubules and loss of ATPase activity. Similar mutations within the P2 and P4 motifs caused a reduction in ATPase activity, but did not appear to affect microtubule binding.[480,481]

The stalk domain, as has been stated, is located between the fourth and fifth AAA modules and comprises an anti-parallel coiled-coil structure with a small globular microtubule-binding region that has been shown to be ATP sensitive. This stem region shows very poor conservation between dynein heavy chain members. The stalk binding of microtubules has been observed using recombinant stalk tip peptides, where they were observed to bind tubulin dimers with a periodicity of 8 nm, similar to that observed with kinesin motor proteins.

Recent structural studies on a dynein MTBD fused to a seryl-tRNA synthetase (SRS) have confirmed the coiled-coil structure (comprising CC1 and CC2) of the dynein heavy chain stalk, with the observation of three heptad repeats ended by two proline residues, which create a kink in the structure followed by a novel fold of α-helices (H1–H6) that form the microtubule-binding motif.[482]

Since the CC2 domain interacts directly with the MTBD movement of the CC1 partner, it was proposed that movement of the coiled-coils with respect to each other could cause differential binding to the microtubule and this indeed was shown to be the case.[482] Based on a series of varying constructs of this peptide, where the CC2 region was truncated causing relative repositioning of the two coiled-coil regions, it was observed that regulation of the binding affinity of the MTBD was moderated based on the relative positioning of the two coiled-coil regions with respect to each other.[482] More recent studies have confirmed this observation by mutation of cysteine residues into the coiled-coil regions of the stalk in order that "intermediates" of the stalk could be trapped when in the presence of an oxidising agent.[483]

Based on these experiments, it has been proposed that the dynein heavy chain motor domain directly controls the state of the MTBD through the coiled-coil stalk. Primary sequence analysis indicates there is a possibility that the sliding of the outward helix (CC1) and the return helix (CC2) may involve a half-heptad shift, based on the sequence of hydrophobic residues on the CC1 helix.

4.3.11 Mechanism

The lack of an atomic structure for dynein heavy chain makes the prediction of a structural model that explains motility difficult; however, single-molecule experiments suggest that dynein complexes walk in a hand-over-hand manner, with more irregular and different stepping behaviour to that of kinesin.

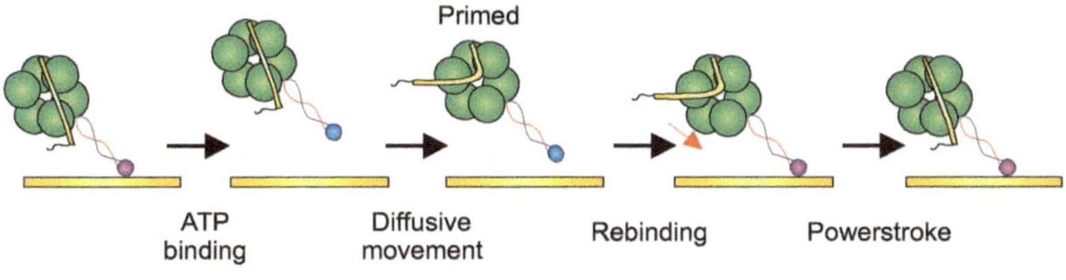

Figure 4.13 Mechanism of dynein movement. Illustration representation of the ATP hydrolysis cycle of the dynein motor. Upon binding of ATP, the dynein molecule is released from the microtubule, followed by a re-cocking of the linker domain. The rebinding of the dynein stalk to the microtubule is then followed by a displacement of the linker domain (giving rise to the power stroke), which generates the movement of the dynein with respect to the microtubule. Adapted from reference [484].

Cytoplasmic dynein has been shown to translocate microtubules via coordination of its two motor domains; however, the specific details of step size (4–32 nm), stall force, and directionality have yet to be agreed.[485] There appears to be some evidence that the small linker region, which joins the tail and motor regions, facilitates dynein motility as it has been observed to move its position relative to the motor domain ring during ATP hydrolysis.[474] A proposed mechanism for dynein motor functions, follows the illustration depicted in Fig. 4.13, where ATP binding causes release of the MTBD from the microtubule,[478] followed by diffusive movement of the dynein molecule along the microtubule during which time a proposed "re-cocking" of the linker (neck) domain occurs. Rebinding of the MTBD to the microtubule results in displacement of the linker region (power stroke), which generates further movement relative to the microtubule.

4.4 Helicases and Translocases

4.4.1 Function

Helicases are a family of proteins whose purpose is to separate duplex DNA, RNA, and occasionally heteroduplex DNA/RNA into separate strands through translocation, and they are essential in a multitude of cell events, including DNA replication, transcription, translation, recombination, DNA repair, and ribosome biogenesis. Helicases require ATP for activity in order to break the hydrogen bonds between each paired base and have been shown to work actively to force the strands apart.

One interesting capability of some helicases is that translocation results in the removal of DNA- or RNA-binding proteins[486] and this capability may have important applications in bionanotechnology. The Type I restriction-modification enzymes and RecBCD have been observed to promote protein displacement[487] for proteins such as the *lac* repressor and even nucleosomes (unpublished observations and reference [486]). Recent studies of DEAD-box proteins from *E. coli* has shown that these helicases are capable of RNA remodelling in an ATP-independent manner.[488]

4.4.2 Classification

A huge number of helicases from a great number of species have been identified (e.g., at least 14 helicases have been confirmed in *E. coli* and 24 in human cells) reflecting the great variety of processes in which strand separation is required. Helicases were originally divided into five superfamilies by Gorbalenya and Koonin[192] based on primary structure analysis:

- **Superfamily I**: UvrD (*E. coli*, DNA repair), Rep (*E. coli*, DNA replication), PcrA (*Staphylococcus aureus*, recombination), Dda (bacteriophage T4, replication initiation), RecD (*E. coli*, recombinational repair), TraI (F-plasmid, conjugative DNA transfer).
- **Superfamily II**: RecQ (*E. coli*, DNA repair), eIF4A (Baker's Yeast, RNA translation), WRN (human, DNA repair), NS3 (Hepatitis C virus, replication). TRCF (Mfd) (*E. coli*, transcription-repair coupling).
- **Superfamily III**: LTag (Simian Virus 40, replication), E1 (human papillomavirus, replication), Rep (Adeno-Associated Virus, replication, viral integration, virion packaging).
- **DnaB-like family**: DnaB (*E. coli*, replication), gp41 (bacteriophage T4, DNA replication), T7gp4 (bacteriophage T7, DNA replication).
- **Rho-like family**: Rho (*E. coli*, transcription termination).

It is important to note that this classification system did not include all putative helicases. As a result, a recent review by Singleton *et al.*[4] proposed the introduction of three further superfamilies to encompass the DnaB-like family (superfamily IV), the Rho-like family (superfamily V), and also a separate family to incorporate further helicases and translocases that are members of the AAA$^+$ proteins[489] (ATPases associated with various cellular activities): Superfamily VI. In addition, the authors also proposed a nomenclature for describing the direction of the enzyme activity (**A** denoting a 3′–5′ directionality of the helicase or **B** denoting a 5′–3′ directionality) as well as substrate specificity (Type α denoting a preference for single-stranded nucleic acid or Type β denoting a preference for double-stranded nucleic acid).

There are several core motifs that all known helicases exhibit. Every helicase has two regions of conserved residues that are located between two RecA-like domains, related to the Walker A and B motifs,[490] required for NTP binding and hydrolysis and one region of conserved residues corresponding to an "arginine finger", which has been shown to be essential for coupling of the energy to conformational change of the proteins.[491] Beyond these core motifs, however, there are a number of motifs that are specific to each family that aid with their classification into their relative superfamily (Fig. 4.14).

4.4.3 Superfamily I (SF1)

All SF1 helicases characterised to date have been shown to be true helicases and show a preference for single-stranded nucleic acid as a substrate (Type α). The superfamily can be further divided, however, based on the direction of translocation: members of the SF1 superfamily can translocate the DNA in either 3′–5′ (A) or 5′–3′ (B) direction.[4]

Structurally, the SFI helicases have largely been characterised as monomeric, although activity has been shown to be enhanced by cooperation with other monomers.[492–495] All members have an ATP-binding site (motifs 1 and 2) located in the cleft between two recA-like domains, and the Walker B motif (2) of the SF1 superfamily can be identified by the distinct DE(Y/F/A)QD-conserved sequence.[496] Two additional motifs, 3 and 6, have been shown to be required for coupling of ATPase and helicase activity, and the motifs 1a,[497] TxGx,[498] 4, 4a,[499] and 5 have all been implicated as being directly involved in handling the bound nucleic acid substrate through its phosphodiester backbone,[192,500] although some of these motifs are superfamily and/or subfamily specific.

4.4.3.1 SF1A helicases

The most studied SF1A enzymes remain UvrD (*E. coli*),[501] Rep (*E. coli*),[502,503] and PcrA (*Staphylococcus aureus*).[504] The crystal structures of the PcrA and Rep proteins both reveal a similar tertiary structure comprising two domains, each of which is further split into two sub-domains by two functionally related domains (in the case of PcrA domains 1B and 2B that are utilised for interactions with the DNA substrate).[505,506]

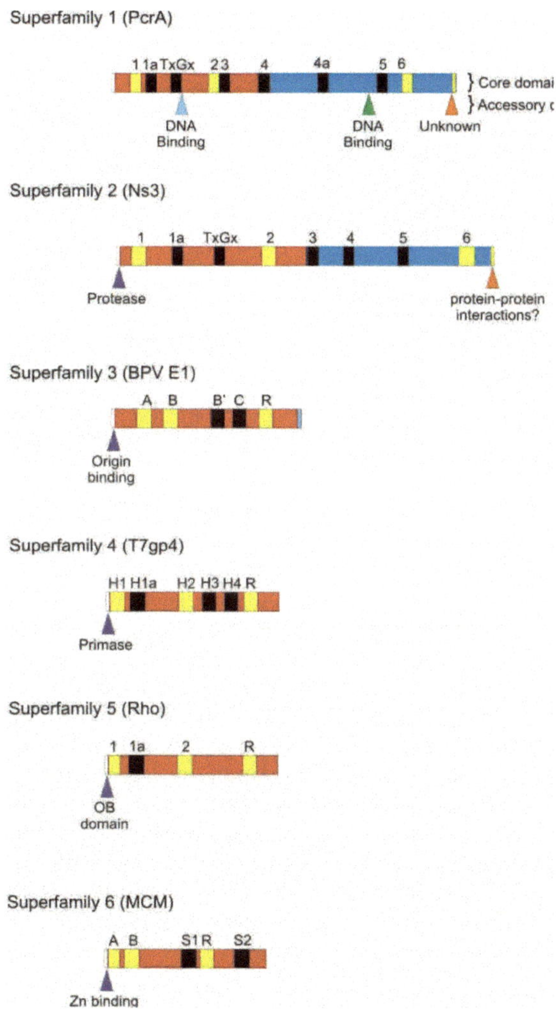

Figure 4.14 The domain structure of the helicase superfamilies. A bar representation of the domain structure of the subfamilies of typical helicases showing the core domains (red and blue), the universal structural elements (yellow), and the superfamily specific structural elements (black). Underneath are noted the positions of accessory domains that are inserted within the two RecA-like folds. Adapted from Singleton *et al.*[4]

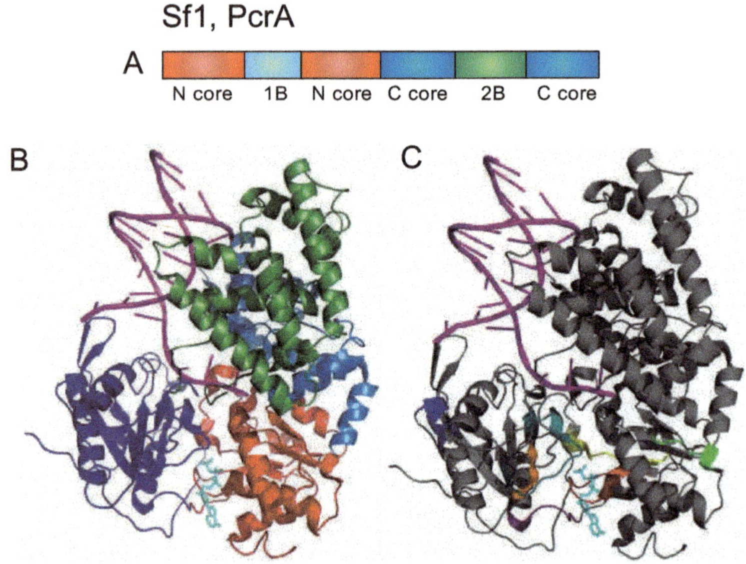

Figure 4.15 The structure of the PcrA helicase of superfamily 1. (A) A bar representation of the domain structure of PcrA showing the core domains (1A and 2A) in red and blue, and the functionally related accessory domains (1B and 2B) in pale blue and green; (B and C) Crystal structure of PcrA bound to dsDNA and ATP (Rendered using PyMol v0.99, PDB code 3PJR). Both (B) and (C) show the bound dsDNA in magenta and the ATP ligand in cyan. Render (B) shows the N and C core domains 1A and 2A (red and blue, respectively). The corresponding accessory domains sub-domain 2B (yellow), and the sub-domain 1B (green) are also shown. Render C shows the arrangement of the conserved motifs[500] surrounding the ATP-binding cleft; motif I (red), motif Ia (green), motif II (blue), motif III (yellow), motif IV (deep purple), motif V (deep teal), and motif VI (orange). Adapted from reference [4].

In case of the PcrA helicase structure, the ATP-binding site (Walker A and Walker B motifs) is located in the cleft between the two RecA-like folds (domains 1A and 2A), which also harbours the majority of the other SF1 accessory motifs (1a, TxGx, 3, 4, 4a, and 5) required for the coupling of ATP hydrolysis to movement and for ssDNA binding (Fig. 4.15). The initial structure for Rep was solved with a 16 mer ssDNA allowing for further information regarding the interactions that are important for ssDNA binding and associated conformational changes within the protein.[506] Further structures of PcrA complexed with a DNA substrate and either a non-hydrolysable analogue of ATP, or a sulphate ion, also yielded further conformational states.[507] This structural information combined with supporting biochemical data[508–512] allowed the development of a model for the duplex unwinding by PcrA and tentatively all SF1A members[513,514] and is described briefly below.

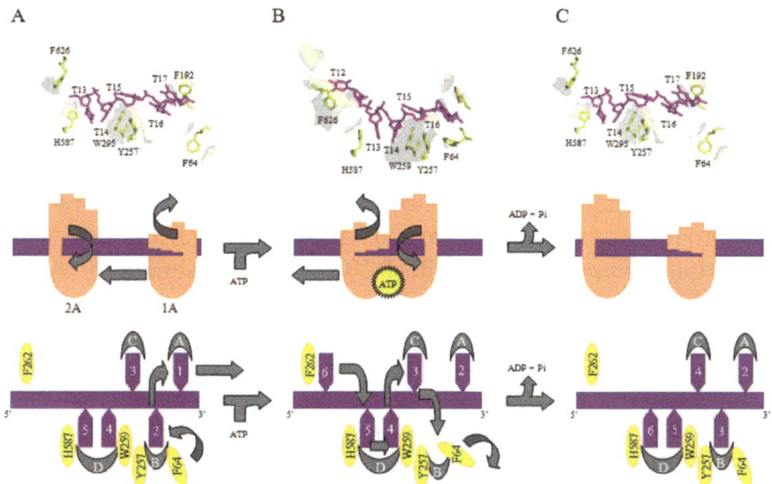

Figure 4.16 The proposed model for translocation by superfamily 1 helicases. Based on the intermediate crystal structures of Rep[506] and PcrA[505] with bound ssDNA during alternative ATP hydrolysis states. (A) to (C) show the three translocation steps of PcrA associated with ATP binding and hydrolysis. The top row shows the stacking between the amino acid residues (yellow) and the ssDNA bases (magenta). The middle row is an analogy of each stage; the gripping hands depict which domains have tight or loose grips on the bound ssDNA. The bottom row shows an illustration of the structural data in the top row, with the bases (1–6) switching between the enzymes nucleotide binding pockets labelled A-D. Adapted from Singleton *et al.*[4]

The proposed model for duplex unwinding (Fig. 4.16) is initiated by an interaction between the helicase and the ssDNA,[501] inducing a conformational change that involves a rigid body movement of sub-domain 2B allowing duplex DNA to interact with a new interface on the surface of sub-domains 1B and 2B. When an ATP molecule is then bound, it is proposed that the cleft between the 1A and 2A domains closes around the DNA molecule causing movement of the 1B and 2B sub-domains in such a way that the bound duplex DNA is forced onto a negatively charged surface, destabilising base pairing between the strands and causing localised denaturation. Additionally the closing of the cleft forces bases of the ssDNA to flip between pockets in the surface of the 1A and 2A motor domains giving rise to the translocation of the helicase along the ssDNA substrate.

4.4.3.2 SF1B helicases

The most studied of the SF1B subfamily are the bacterial RecD and Dda helicases, although eukaryotic forms, such as Pif1 and Rrm3, have been studied to a lesser extent.[515] The only available crystal structure to date is that of RecD,[516] although it is important to note that this is by no means a standalone enzyme and is an important component of the RecBCD complex (essential for repair of double-stranded breaks) on which the majority of biochemical characterisation has been focused. The Dda helicase has been extensively characterised[493,517–521]; however, no structure has been solved for the enzyme and, therefore, information about the mechanism of the SF1B helicases has mainly been predicted based on an amalgamation of the data available on both proteins.

It has been shown that both enzymes are functional as monomers,[522,523] despite the fact that RecD is part of heterotrimer. Structurally, RecD shares features with SF1A enzymes, with two RecA-like domains forming an ATP-binding site between them. The mechanism of translocation has also been proposed to be similar to the SF1A enzymes, with an inchworm-like mechanism; however, the precise mechanism and residues involved cannot be predicted due to the lack of conservation between the two subfamilies important residues in motif 3 and the Phe residue in motif 1a are not conserved in the SF1B helicases).[4]

4.4.4 Superfamily 2 (SF2)

Helicase superfamily 2 contains the greatest number of characterised helicases, including the DEAD-box RNA helicase,[524] the RecQ-like family,[525] and the Snf2-like enzyme subfamilies.[526,527] The basic function of an SF2 family member is probably an ATP-dependent translocation on either ssDNA or dsDNA, although some members have been implicated as non-processive mechanical switches that cause alteration of nucleic acid and nucleo-protein conformation.[528]

In contrast to the SF1, it appears to be possible to subdivide the superfamily based on both the direction of translocation (α and β) as well as the substrate for translocation (A and B) based on those members characterised to date. The majority of the biochemically characterised members of SF2 exhibit a type A polarity (3′–5′); however, there are exceptions, including the XPD family,[529] as well as some members (DEAD box-related RNA helicases) that have been shown to weakly catalyse unwinding in both directions in a non-processive manner.[530] Due to this, these bidirectional enzymes have been proposed as remodelling switches rather than *bona fide* translocases, but in some cases it has been shown that the processivity of the enzyme is stimulated with additional protein cofactors.[4]

The majority of SF2 enzymes have accessory domains as inserts within their sequence, or additional sequence at either the N-terminus or C-terminus, of the core motor domain. Generally, these additional domains appear to be involved in the mechanism of the helicase (specific substrate targeting); however, in some cases the domains have related enzyme activity, such as protease (NS3), or endonuclease domains (from Type I restriction-modification enzymes).[4]

4.4.4.1 SF2α subfamilies

Of the SF2 members, non-structural protein 3 (NS3) of the hepatitis C virus (HCV) is probably the best characterised (due to its central role in the RNA replication of HCV), which is capable of unwinding DNA, RNA, and hybrid duplexes[531] with a 3′–5′ overhang[532] (marking it as a member of the SF2Aα) and this movement has recently been tracked allowing modelling of the mechanism of translocation.[533] The enzyme has been shown to function as a monomer[534]; however, several studies have also implicated a functional NS3 dimer.[495,535] An analysis of the domain structure (Fig. 4.17) shows the presence of the two Walker A and B motifs (1 and 2), motifs 3 and 6 responsible for coupling of ATP hydrolysis to translocation activity, and motifs 1a, TxGx, 4, and 5 that are all thought to be involved in substrate binding. It is important to note that motif 3 of SF2 helicases does not share sequence homology with that of the SF1 motif 3 and is located in a different part of the enzyme.[536]

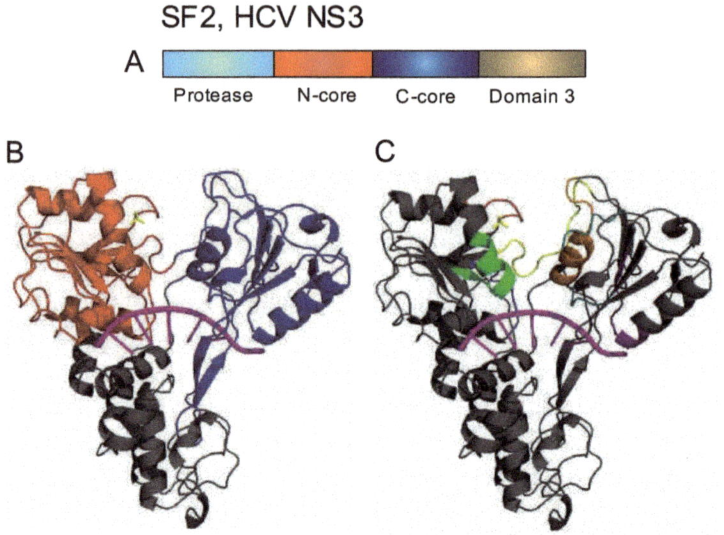

Figure 4.17 The structure of members of the SF2 family of helicases. (A) A bar representation of the domain structure of PcrA showing the four domains of the NS3 helicase; (B and C) tertiary structure of HCV NS3 complexed with ssDNA (rendered using PyMol v0.99, PDB code 1A1V). Both renders (B) and (C) show the bound ssDNA in magenta, and the sulphate ion, marking the ATP-binding site, is shown as a stick-model at the top (yellow). Render (B) shows the N- and C-core domains (red and blue, respectively). Render (C) shows the arrangement of the conserved motifs[500] surrounding the ATP-binding cleft; motif I (red), motif Ia (green), motif II (blue), motif III (yellow), motif IV (deep purple), motif V (deep teal), and motif VI (orange). Adapted from reference [4].

Crystallography of the helicase domain of NS3 alone and in complex with single-stranded nucleic acid has yielded useful structures that give insight into the mechanism by which this helicase translocates DNA (Fig. 4.17). Similar, to the SF1 helicases, the core of the enzyme comprises two RecA-like core domains, which sandwich the nucleotide-binding pocket that contains the majority of the helicase motifs and part of the ssDNA-binding site. The crystal structure does not include the protease domain of NS3 (required for processing of the viral polyprotein), but does include an accessory domain that prevents re-annealing of ssDNA.[537]

The ssDNA co-crystallised with the NS3 was observed to run along the top of the RecA-like domains with six bases in direct contact with this region, in a similar position to that observed in the structure of PcrA; however, the way in which the bases are bound differs significantly. In the PcrA structure, the ssDNA is held by base stacking interactions with aromatic residues, while the ssDNA in the NS3 structure adopts a pseudo B form structure with stacking between the bases maintained. Additionally, two residues (V432 and W501), termed "bookend residues", that have been determined to be important by mutagenesis[538–540] were seen to sandwich five of the bases across the motor domains. The regions around the bookend residues form equivalent DNA-binding regions, which are believed to be involved in a ratchet-like translocation mechanism.[536]

The mechanism proposed relies upon the closing and opening of these domains dependent on ATP binding and hydrolysis, coordinated with the gripping and release of the DNA substrate, and indeed this model is made more credible by a structure of a related RNA helicase, VASA, which was co-crystallised with a non-hydrolysable ATP analogue and yielded a closed complex.[541] The mechanism proposed is similar to that suggested for PcrA, and indeed the only significant difference between the two mechanisms is the way in which the two types of helicase grip the substrate DNA.

With NS3, the kinetics of the unwinding mechanism have been investigated, and the enzyme has been shown to have a large step size (9 bp, 11 bp, or 19 bp in size)[495,534,542]; whereas, the related NPH-II helicase appears to unwind in 6 bp steps.[543] With regard to substrate translocation, NS3 and related helicase are unable to translocate across gaps in the substrate strand,[544,545] but are able to translocate long strands of polyglycol linkers in RNA and kinetic data on this substrate confirms the largest step size (18 bp).[545] ssDNA binding by NS3 is weaker in the presence of non-hydrolysable ATP analogues, and the enzyme is capable of melting several nucleotides of the duplex simply through binding energy.[546,547] These observations provide a basis for a "Brownian ratchet" translocation mechanism whereby the binding of the ATP molecule causes a weakening of the binding of the DNA by the enzyme, allowing a period during which Brownian motion causes the enzyme to move on the strand before hydrolysis of the ATP molecule causes the enzyme to once again clamp onto the substrate DNA with a bias towards movement forward on the strand.[547]

4.4.4.2 SF2β subfamily

Characterisation of the translocation of enzymes on dsDNA without unwinding of the substrate is relatively rare in the literature, due to the difficulty in assaying for DNA tracking, as there is no discernable product. In order to characterise these events, however, experiments involving triplex displacement have been used to demonstrate the translocation of dsDNA across several proteins, the original of which was EcoR124I (a Type IC restriction endonuclease).[548] Only one such SF2β structure has as yet been co-crystallised with duplex DNA, namely, Rad54 from *S. solfataricus* (Fig. 4.18).

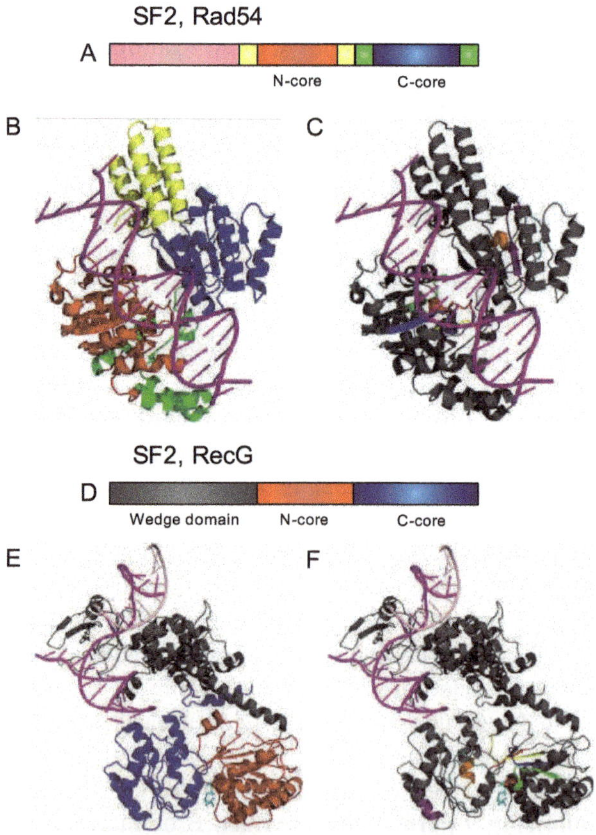

Figure 4.18 The structure of the Rad54 and RecG helicases. (A) Primary structure analysis of the Rad54 protein, showing the relative positions of domains within the sequence; (B and C) crystal structure of Rad54 with Bound dsDNA (rendered using PyMol

Figure 4.18 (*Continued*)

v0.99, PDB code 1Z63). Both (B) and (C) show the bound dsDNA in magenta. (B) shows the N- and C-core domains (red and blue, respectively) and the corresponding sub-domains, 2B (yellow) and 1B (green), are shown at the top and bottom correspondingly. (C) shows most of conserved motifs[500] arranged within the N-core (1A) domain; motif I (red), motif Ia (green), motif II (blue), motif III (yellow), motif IV (deep purple), motif V (deep teal), and motif VI (orange). (D) Primary structure analysis of the RecG protein, showing the relative positions of domains within the sequence. (E and F) The crystal structure of RecG bound at three-way DNA junction[551] (rendered using PyMol v0.99, PDB code 1GM5); demonstrates how the N-terminal accessory wedge (grey) acts as a specificity factor for the three-way DNA junction. (E) shows the DNA duplex (magenta) coordinated towards the N- and C-core domains (red and blue, respectively), in a similar fashion to that of Rad54. The lagging nascent strand is represented in light pink. The ATP (cyan) binding site is shown situated in the cleft between 1A and 2A. (F) shows most of the conserved motifs[500] arranged within the N-core (1A) domain; motif I (red), motif Ia (green), motif II (blue), motif III (yellow), motif IV (deep purple), motif V (deep teal), and motif VI (orange). Adapted from bibliography reference.[4]

As with the other crystal structures reviewed so far, the tandem RecA-like folds are the predominant feature of the structure of Rad54; however, in this enzyme, each of the RecA-like domains is modified with the addition of two insertions of unknown function, and the C-core domain is rotated ~180° from its canonical position. Comparison with the Rad54 structure from zebra fish reveals a C-core domain of the motor that is maintained in an "open" position implying that the domain may undergo conformational change to this position before translocation can begin.[549]

The Rad54 structure from *S. solfataricus* reveals that interactions with the dsDNA are mainly through the N-core domain involving motifs 1a and also Snf2 subfamily specific motifs that interact with the phosphodiester backbone of the substrate along the minor groove. In contrast, the interactions of the C-core with the DNA appear to be minimal, possibly due to the relative orientation of the domains.[550] This substrate binding is similar to that found in the NS3 helicase.

These observations imply that the SF2β subfamily members may conform to a model whereby the main motor-active contacts of the translocases are with a single strand of the duplex, with an inchworm mechanism of movement similar to those proposed for both NS3 and PcrA. This model would result in the dsDNA substrate twisting across the surface of the motor domains, without duplex unwinding. In case of EcoR124I, where the motor subunit HsdR is part of a pentameric complex with one of the subunits responsible for substrate recognition and binding, the motor activity would result in the generation of torque.[4]

4.4.5 Chromatin Remodelling Complexes

ATP-dependent chromatin remodelling complexes reposition histones complexes on DNA using a DNA-translocation mechanism, and these motors have been classified as members of the SF2 family of helicases based on conserved protein motifs.[552] The forces required to disrupt the nucleosome structure are between 10 and 30 pN,[553] and the magnitude of the driving force of the chromatin remodelling complex will dictate the mechanism of nucleosome disruption, with low forces likely to lead to loop or wave formation, which then propagates around the histone surface,[554,555] or weaken DNA-histone interactions upon binding nucleosomes.[556,557] While larger driving forces allow a much more diverse mechanism for remodelling because many more DNA-histone interaction can be disrupted in a more concerted manner.[558] Recently, single-molecule measurements with a minimal chromatin remodelling complex has shown a very high driving force of 30 pN,[559] which suggests a highly disruptive mechanism. It is interesting to note that another member of the SF2 family, EcoR124I, has also been shown to displace a variety of proteins from DNA[487] and this motor, which generates forces up to 8 pN, has also been shown to displace histones from DNA (unpublished observations). However, while ATP-dependent initiation of translocation by EcoR124I has been shown to require localised melting and distortion of DNA,[560] the minimal chromatin remodelling complex appears to require a reiterative short translocation mechanism to initiate translocation (reminiscent of abortive transcription).[559] The motor translocates DNA at 25 bp s^{-1} with a step size of 2 bp and a processivity of 35 bp,[559] indicating that the motor is not unlike other helicase-like motors.

4.4.6 Superfamilies 3–6

The last four helicase superfamilies proposed are made up of the hexameric helicases. These encompass superfamily 3, DnaB -like helicase (family 4), Rho/V-F-ATPases (family 5), and the AAA$^+$-like family (superfamily 6), which for simplicity were proposed to be renamed to superfamilies 3–6[4] (Fig. 4.14).

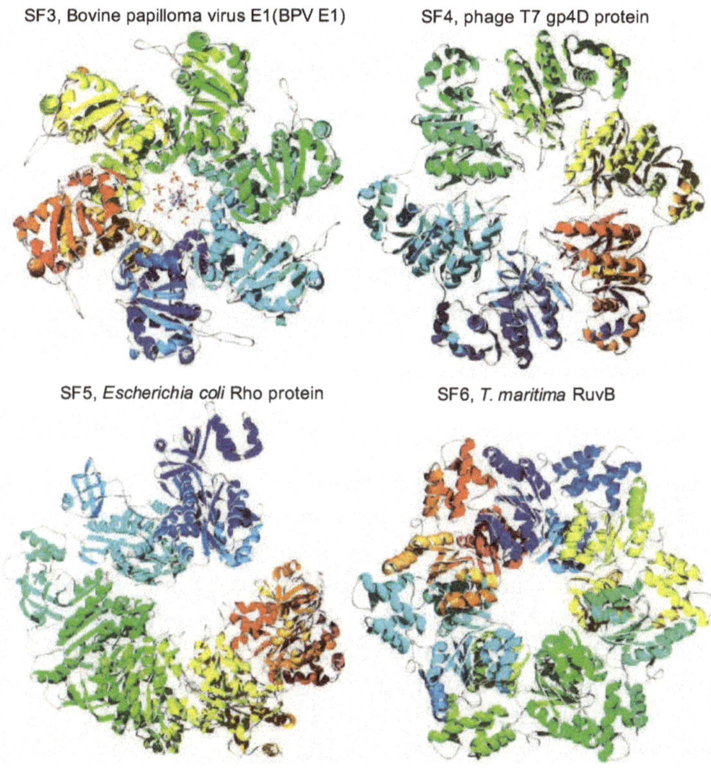

Figure 4.19 Structure of the hexameric helicases. Ribbon representations of hexameric helicases from SF3-6. Adapted from Singleton *et al.*[4]

There is remarkable similarity between these superfamilies (Fig. 4.19), which structurally all resemble hexameric toroids. In contrast to SF1 and SF2 members, the ATPase core is located at the interface between monomers rather than within a single polypeptide. The major distinguishing factor between these superfamilies is the type of fold for the ATPase core, where either a RecA-like or AAA$^+$-like fold is found. The nucleotide-binding pockets usually have an arginine finger involved in coupling of ATP hydrolysis to translocation activity.

Most of the hexameric helicases have roles in DNA replication, recombination, and transcription, for which they must be capable of unwinding the DNA substrate over large distances. In order for the helicase activity to occur, the doughnut-shaped hexamer must first be loaded onto the DNA, and there is evidence that some can simply self-assemble around the substrate, whereas others have been shown to require loading partners for complex assembly.[561]

4.4.6.1 Superfamily 3

Originally identified in small genome DNA and RNA viruses,[562,563] the SF3 helicases have been determined to have an origin-binding domain located at the N-terminus of a variant AAA$^+$ core that contains nucleotide-binding motifs and a set of loops involved in DNA binding or protein–protein interactions.[564]

The conserved motifs of SF3 correspond to the Walker A and B domains (A and B), the B′ motif that is responsible for interactions with the DNA substrate, and the SF3 specific C motif, after which is found the conserved arginine finger.

Hexamer formation appears to be stabilised via a less well-conserved helical domain located N-terminally to the motor core. In the structure of the bovine papilloma virus E1 helicase, this oligomerisation domain forms a rigid collar stabilising the hexameric structure. This particular SF3 helicase was co-crystallised with ssDNA, which was found enclosed within a ~13 Å channel at the centre of the hexamer, with the loop regions of the B′ motif extended and interacting with the phosphodiester backbone in a spiral orientation.[565]

Although nearly all of the nucleotide-binding sites of each monomer were occupied by an ADP molecule, the relative positions of the associated loop from the B′ motif (P-loop) were found in one of three different conformations. These were interpreted as ATP binding, ADP binding, and empty conformations (in a manner similar to that observed with ATP synthase), and each monomer had decreasing levels of interaction with the ssDNA. The ATP-type sites had the loop orientated towards the 5′ end of the ssDNA, the ADP-type at an intermediary position, whilst the empty sites loops were orientated towards the 3′ end of the substrate.

This orientation of the DNA-binding loops in the structure allowed the authors to propose a mechanism whereby a sequential ATP binding in each monomer causes movement of the loop and individual bound nucleotide of the ssDNA substrate downwards through the channel, followed by a release following dissociation of the ADP. This mechanism would require coordination between the six subunits presumably through interactions directly between monomers.

4.4.6.2 Superfamily 4

The SF4 members were initially identified in bacteria and bacteriophages as being helicases involved in DNA replication. In bacteria, the helicase associates with a primase monomer, whereas in bacteriophage, the helicase and primase activities are retained within the same protein. All SF4 helicases characterised have a 5′–3′ (type B) polarity.[4]

The conserved motifs of the SF4 family, H1 and H2, are equivalent to the Walker A and B motifs, whereas motifs H1a, H3, and H4 are unique to this superfamily (Fig. 4.14). The H3 motif, however, encompasses a conserved glutamine (or occasionally histidine) residue, which fulfils the role of the motif 3 found in SF1. The Walker B motif contains an isolated aspartate embedded within a hydrophobic stretch of amino acids as opposed to the DExx motifs of the SF1 and 2 enzymes.[566]

Of the SF4 helicases, the bacteriophage T7 gene 4 protein is probably the best characterised and comprises a single polypeptide with both a primase and a helicase activity.[567] Structures have been solved for two constructs of the gp4 protein — gp4E and gp4D. The gp4E construct is a truncated form of the gp4 protein that retains NTPase activity and this crystallised to provide a monomeric helical filament resembling the core of a RecA-type fold.[568] The gp4D construct encompasses the whole of the helicase domain and when crystallised yielded a hexameric ring structure.[569]

Analysis of these structures reveals that motif H4 forms a loop that projects into the centre of the channel within the hexamer, and this has been implicated in ssDNA interactions by mutational analysis.[570] The structure, co-crystallised with a non-hydrolysable analogue of ATP (ADPPNP), showed nucleotide binding at four out of six of the ATPase sites at the subunit interfaces. Interestingly, the overall structure resembled an asymmetric hexamer, with three different orientations for the monomers, which were proposed to reflect their relative binding competency: NTP-binding competent (T), NDP competent (D), and empty (E) states. The structures without a bound nucleotide or substrate are observed as a symmetric hexamer, and it was proposed that the asymmetry of units of the gp4D structure suggested a link between ATP hydrolysis and DNA translocation, leading the authors to propose a model for a sequential hydrolysis of ATP (discussed later).[569]

4.4.6.3 Superfamily 5

The Rho helicase is essential for factor-dependent transcription termination and is essentially very closely related to the SF4 family, although it was originally placed in a separate family based on primary structure. The Rho helicase functions by binding to a specific target sequence and then unwinding the DNA–RNA hybrid in an ATP-dependent manner, causing the release of the RNA from a transcribing elongation complex.

Analysis of the primary structure reveals two motifs corresponding to the Walker A and B motifs: the R motif corresponding to a conserved arginine residue, and the family specific 1a motif (Fig. 4.14).

Both electron microscopy and crystallography of Rho yielded "open" and "closed" forms of the hexamer. In the electron micrographs, the "open" forms were proposed as either an open "notched form" of the hexamer or a pentamer,[571] whereas the crystallographic structures displayed an open "lock washer" conformation, with a left-handed spiral conformation of the monomers, which gave an opening sufficiently wide (12 Å) to allow for the passage of ss-RNA.[572] Subsequently, a structure has been obtained with ssRNA bound across the motor domains, and this displays a closed, flat conformation, indicating that the "open" form of Rho may be a loading intermediate.[573]

4.4.6.4 Superfamily 6

Members for SF6 were proposed to be hexameric helicases that display an AAA$^+$ fold, but do not fall into the SF3[4] and one example of these is the mini chromosome maintenance (MCM) complex, which is essential for replication initiation and elongation.[574]

In eukaryotes, this hexameric complex comprises six different subunits (MCM2-7) each of which has a highly conserved helicase domain (MCM core domain), with the conserved Walker A and B motifs (A and B) and arginine finger motif (R), as well as two SF6 specific motifs S1 and S2. At the N-terminus of four out of the six eukaryotic MCM members, a Zn^{2+} binding motif is observed and this has been shown to be required for helicase activity in archaeal MCM, which are generally agreed to be homohexamers.[575]

Separately, MCMs 4, 6, and 7 have all been shown to exhibit weak helicase activity *in vitro*; however, the complete hexamer does not show helicase activity unless in complex with additional peptides (cdc45 and GINS).[576]

Although no crystal structures have as yet been published, cyro-EM models of an intact archaeal MCM confirm the oligomerisation of the subunits to form a hexameric ring, with a large channel running through the centre that would be large enough to accommodate dsDNA. A model based on the mitochondrial MCM domain structure and the AAA^+ domain of RuvB was shown to fit into the electron density map provided.[577]

RuvB is another AAA^+ protein that does not conform to the SF3 based on primary structure and is responsible, alongside RuvA and RuvC, for processing Holliday junctions, intermediates in homologous recombination.[578] The RuvB protein is ATP dependent and responsible for migrating Holliday junctions through the hexameric structure. Although there are no structures available for the hexameric form of prokaryotic RuvB, the crystal structure of the homolog from humans, RuvBL, crystallised as a hexamer, with an asymmetric arrangement of monomers, consistent with observations for other hexameric helicases and the central channel was observed to be large enough in diameter to accommodate both dsDNA and ssDNA substrate.[579]

4.4.7 Other Hexameric Translocases

There are a number of proteins that have what have been described as helicase core motifs (i.e., Walker A and B motifs and a conserved arginine finger domain that are considered essential for helicase activity), that use dsDNA as a substrate, yet do not conform to the standard helicase description; a protein capable of unwinding double-stranded nucleic acids. Essentially, all of the helicase superfamilies described are translocases, in that they translocate a double strand of DNA, and it appears to be mainly through interaction with accessory domains, helicase motifs, or structure-induced occlusion that the displacement of a single strand occurs.

These other nucleic acid translocases share a great deal of homology with their helicase counterparts but lack the motifs required for strand separation. The hexameric protein FtsK is one such protein, involved in segregation of daughter chromosomes during vegetative cell division.[580] The protein comprises an N-terminal domain that acts to localise the protein to the division septum[581,582] and a C-terminal domain that provides the translocase activity,[583] but multiple regions have been implicated in the final stages of chromosome segregation.[584] The protein functions by translocating DNA until it reaches a dimer resolution site where activation of site-specific recombinases occurs.[585]

The crystal structure of the C-terminal motor domain has been solved for both *E. coli* and *P. aeruginosa*. The *E. coli* $FtsK_c$ protein crystallised as a monomer, exhibiting a typical RecA-like fold; whereas the *P. aeruginosa* $Ftsk_c$ crystallised as a hexamer. The RecA-like fold observed in FtsK was considered significantly different to other hexameric helicases in the way in which the fold was formed, with all of the residues for the active site contributed by the same monomer rather than a contribution from neighbouring monomers as we have discussed.

By comparison of the two structures, it was proposed that each monomer had an N-terminal "jaw" domain that is implicated in pulling dsDNA through the core of the hexamer, in an NTPase-dependent manner, postulated to be coordinated through a "rotary inchworm" mechanism, where all subunit sequentially enable movement of the DNA through the bound complex.[586] The FtsK regulatory γ-sub-domain, of the C-terminal domain, determines directionality of translocation through its interaction with specific eight base pair chromosomal sequences (KOPS), which are used to initiate new translocation events rather than reorient existing events.[587]

A structurally related hexameric single-stranded translocase, TrwB, exhibits a similar conformation of the RecA-like fold with active site residues contributed by one monomer of the hexamer.

Similarly to FtsK, the TrwB protein is effectively a DNA pump, mediating DNA conjugation between *E. coli* by acting as a transmembrane pore.[588] The channel through the centre of the hexamer is ~20 Å in diameter, narrowing to ~8 Å, where it encounters an asparagine plug at the cytoplasmic side of the channel hypothesised to be part of a closed conformation of the enzyme that could open upon activation.[589]

Other related enzymes are found involved in viral RNA genome packaging. Phage ϕ12 P4 ATPase is a ssRNA translocase, which is found bound to the viral capsid. Structural studies have revealed a symmetrical hexameric protein complex, with the ATP-binding site located at the interface between monomers and a conical central channel of 21–25 Å in diameter, which is sizeable enough to accommodate the ssRNA substrate.[590]

Although none of these hexameric enzymes unwind DNA and the sequence homology is weak, the structure of their motor domains and potentially the way in which they translocate nucleic acid appears to link them to the SF4 family of hexameric helicases that utilise the RecA-like fold for DNA translocation.[4]

The structural data available for the hexameric helicases discussed up until now have led the respective authors to propose a number of potential mechanisms based upon the structural snapshots obtained through co-crystallisation with NTPs and analogues or by comparison with other similarly structured ATPases (i.e., F_1-ATPase, ClpX, etc.). This has led to five models of NTP hydrolysis and, therefore, the related DNA translocation/helicase activity:

4.4.7.1 F_1-ATPase-like (three-site sequential) model

Based on the perceived similarity between the F_1-ATPase structure and that observed for the SF4/SF5 helicases,[591] a model was proposed by which the hexamer at any given time has three NTP-hydrolysing sites.[592] Each site may exist in one of three states — NTP bound (T), NDP bound (D), or empty (E) — with each one exclusively assigned to each given NTPase site (Fig. 4.20a). The model proposed that each site cycled through a coordinated cycle, whereby cycling from T-E occurred and indeed this model was supported by a number of studies on the T7 DNA helicase. A difficulty with this model is that it was based on F_1-ATPase (that has differing catalytic and non-catalytic chains) rather than a homohexameric system with all sites potentially able to hydrolyse an NTP, and the model has been superseded by an all sites sequential model.[593,594]

4.4.7.2 All sites sequential model

The all sites sequential model expands on the F_1-ATPase site model with the proposal of all sites being active at any given time and a coordinated NTPase activity between adjacent subunits manifested by the appearance of two trimers of subunits (Fig. 4.20b) in the T, D, and E NTP-binding states[569] (or as blocks of binding states,[565] Fig. 4.20c), respectively, each having contact with the nucleic acid substrate at some point during the translocation event. Being such a heavily coordinated and symmetric event, a helicase-hydrolysing NTP and translocating substrate in such a way would be expected to be sensitive to the presence of inactive subunits and indeed experiments have confirmed this to be true with T7gp4 helicase strongly inhibited by the presence of a small number of non-catalytic subunits.[595]

4.4.7.3 Stochastic model

Based on the peptide translocase ClpX (a hexameric AAA^+ family member), a system whereby NTP hydrolysis is effectively uncoordinated with regards to the individual monomers, a model for hexameric helicases whereby ATP hydrolysis can effectively occur at random could be inferred[596] (Fig. 4.20d). As with the all sites sequential model, all sites could potentially be active. This mechanism would mean that inactivation of individual monomers could occur without total inhibition of the translocase activity.

4.4.7.4 Concerted model

The concerted model of DNA translocase activity proposes that every monomer of the hexamer has the same NTP-binding state at any given time with every subunit cycling through the T, D, and E states as a substrate molecule is translocated (Fig. 4.20e). This model has been proposed as representative of the AAA^+ SV40 large T-antigen and potentially other AAA^+ helicases.[597] This mechanism could potentially be related to the stochastic model whereby substrate-binding coordinates this concerted mechanism.

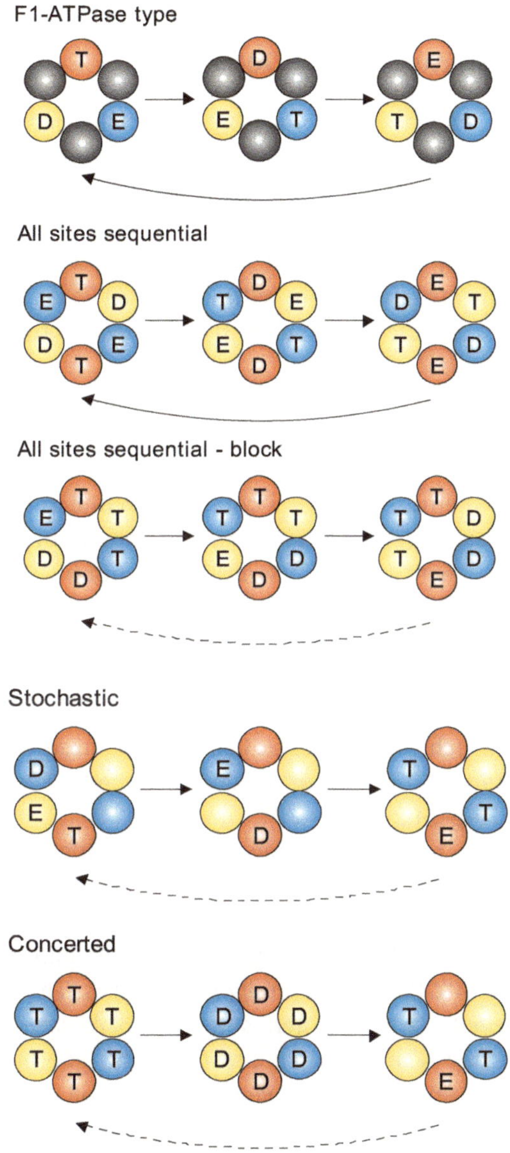

Figure 4.20 A proposed model for NTP hydrolysis within the hexameric helicases. Schematic depicting the proposed models for NTP hydrolysis based upon the conformation of the ATPase site as observed in structural models co-crystallised with NTPs and analogues. Adapted from reference [4].

4.4.8 Structural Homology Between Helicase and Translocase ATPase Domains

Although the sequence homology between helicases can be weak, the available structures for helicases reveal a common architecture that appears to be preserved across helicase superfamilies as well as other prototypical ATP-dependent motors. This has recently been reviewed by Enemark and Joshua-Tor,[598] where comparison of the available structural data with known ATPases has identified individual site residues that have potentially related roles in all translocases.

As has been stated, there are two main folds employed in the helicases to form an active ATPase site: RecA-core fold and AAA-core fold. Despite the topological differences between these two folds, the ATP sites themselves have a common architecture when in an ATP-bound state (referred to as T or "cylinder compressed" states). One side of the active site, present in one subunit, comprises the Walker A and B motifs and a catalytic base of one domain. The adjacent, but facing, subunit provides the other side of the ATP site and contributes two or three basic residues. The inclusion of a separate domain (or even monomer in hexameric motors) in the ATPase site, which contributes at least one of these basic residues, allows for the transmission of a signal between ATPase sites (See Table 4.3 and Figs. 4.21 and 4.22).

Similarities have previously been identified between the structures of PcrA (SF1) and the F_1-ATPase catalytic sites.[600] Enemark and Joshua-Tor[598] further identified significant similarities between these and the whole AAA$^+$ family of proteins as well as the SF2 DEAD-box family of helicases. The conservation of the active site, with the left side of the ATPase site derived from the Walker A and B sites and the right side lined by three basic residues, appears to be similar, based on crystallographic studies, for PcrA,[507] UvrD,[601] F_1-ATPase,[591] NSF,[602,603] Tag,[597] E1,[565] Vasa,[541] and eIF4A3.[604]

Table 4.3 Comparison of the structural homology between helicase ATPase sites and identification of key amino acids and related motifs

Protein		Trigger	Piston	Tether
F_1-ATPase		α-R373, "arginine finger"	β-R189	β-R260
SF1	H3	H4	H6	H5
PerA	Q254	R287	R610	K568
UvrD	Q251	R284	R605	K563
AAA^+	"senor-1"	"senor-2"	"arginine finger"	unnamed
NSF	S647	K708	K631	R607
BPV E1	N523, "sensor-1"	K425, "sensor-2"	R538, "arginine finger"	R493, "sensor-3"
SV40 Tag	N529	K418	R540	R498
SF2 DEAD-box		H6	H6	H5
Hs eIF4A3		R370	R367	R339
Dm Vasa		R582	R579	R551
SF2		H6	H6	H6 + mediator
RecQ		R329	R326	Q322
HCV NS3		R467	R464	Q460
SF4		KxR	KxR	H4 + mediator
T7gp4		R522	R520	Q494
Bst DnaB		R420	R418	R388

Source: Reproduced from reference [599].

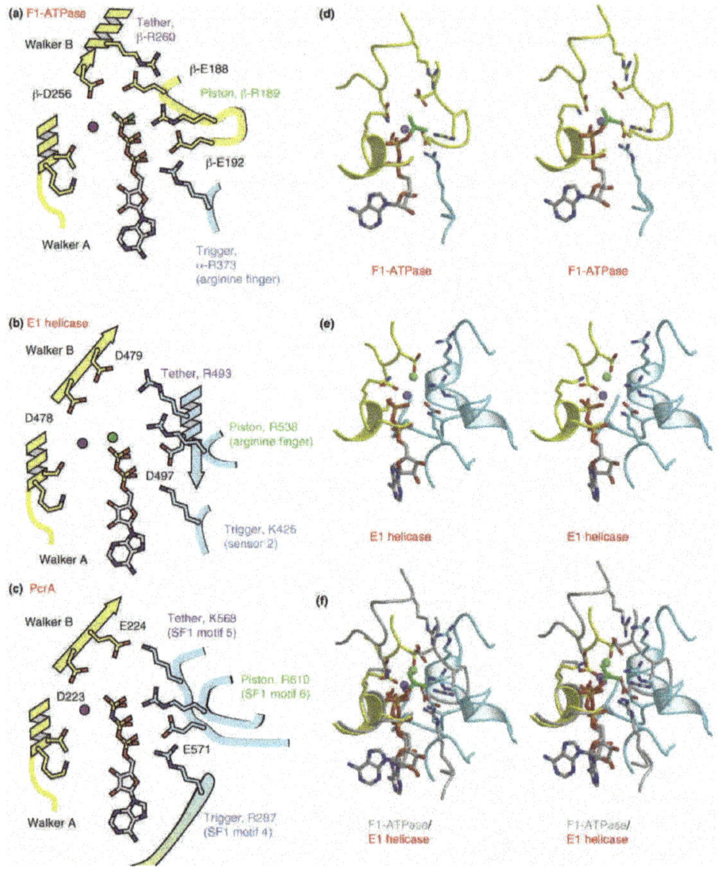

Figure 4.21 Detailed structural features of conserved molecular motor components. Structure of specific examples of the ATP-binding sites of helicases. The Mg^{2+} cation is shown in purple. (A) F_1-ATPase active site with the β-subunit in yellow and the α-subunit in cyan. (B) Papillomavirus E1 with one subunit in yellow and the adjacent subunit in cyan. A chloride ion in the γ-phosphate position is coloured green. (C) PcrA active site with RecA-like domain 1 in yellow and RecA-like domain 2 in cyan. (D and E) Stereo views of the F_1-ATPase configuration bound to ADP-BeF3 (PDB code 1W0J) and the E1 ATP-type configuration (PDB code 2GXA). The subunits are colour coded as above. (F) Structural overlay of the E1 ATP-type configuration with the F_1-ATPase configuration. The E1 structure is in colour, and the F_1-ATPase is in grey. Figure and legend reprinted from reference [599] with permission from Elsevier.

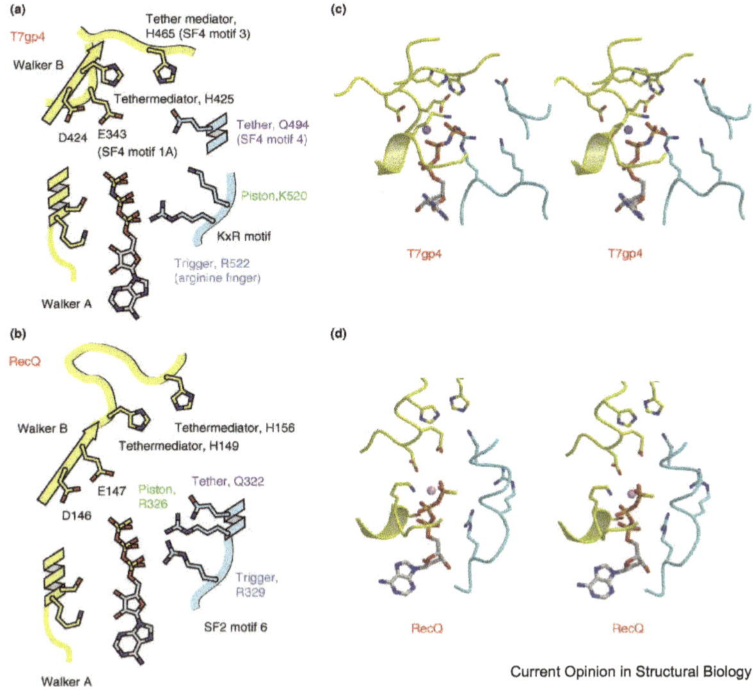

Figure 4.22 Detailed structural features of conserved molecular motor components (continued). The ATP-binding site is tethered between a conserved glutamine residue, a histidine or second glutamine, and often an aromatic residue. (A) Schematic representation of the ATP configuration for T7gp4 with one subunit in yellow and the adjacent subunit in cyan. (B) Schematic of the SF2 helicase RecQ[605] with RecA-like domain 1 in yellow and RecA-like domain 2 in cyan. (C and D) Stereo views of the T7gp4 AMP-PNP (PDB code 1E0J) and the RecQ ATP-γS configurations (PDB code 1OYY). The first tether "mediator" position is generally aromatic (conserved tyrosine in DnaB), and the second position is occupied by either histidine or glutamine (conserved glutamine in DnaB). Figure and legend reprinted from reference [599] with permission from Elsevier.

The first of the conserved basic residues corresponds with the conserved arginine of SF1 helicase motif IV (SF1 H4), the sensor-2 position of AAA^+ enzymes, the arginine finger of F_1-ATPase (α-R373), and the conserved arginine found in motif IV (H6) of the SF2 helicases.

A second conserved basic residue aligns with the conserved arginine of SF1 helicase motif VI (SF1 H6), β-R189 of F_1-ATPase, the arginine finger of AAA^+ enzymes, and a second conserved arginine within the SF2 H6 motif.

The third conserved basic residue corresponds to the SF1 motif (H5), "sensor 3" of E1, β-R260 of F_1-ATPase, and also an arginine within motif V of the SF2 helicase family that is conserved among DEAD-box proteins. This third basic residue appears to have the most flexibility within the three sites, and in structures of SF1 proteins that have an analogue of ATP bound, this residue is mostly a lysine or arginine; but in most SF1 motif H5 sequence alignments, it is most often a glutamine. The same is found for the SF2 helicase alignments where this position is also mainly found to be occupied with a glutamine residue.

4.4.9 A Conserved ATPase Site Mechanism

Based on the arrangement of the active site, the same authors proposed specific roles for each component, comparing each to a component of a standard combustion engine. The first of these three basic residues was labelled as the "trigger" and was proposed to generate the compression in the combustion chamber upon binding of an ATP molecule, differentiating the state of the active site based upon the progression of ATP hydrolysis. In other AAA^+ enzymes, this basic residue is named the sensor-2 residue and is found as part of a larger lid domain, which is believed to be in some way involved in coordinating the reactivity of the ATPase site.[606]

Once ATP hydrolysis has occurred, the second conserved residue (labelled as "piston") must leave the site, which is thought to be coordinated with the loss of an anion from the active site (e.g., post-hydrolysis phosphate). This movement of the piston residues is what has been proposed as causing the conformational change upon completion of ATP hydrolysis.

Whilst inorganic phosphate was proposed to be released upon the movement of the piston, the release of the ADP generated by ATP hydrolysis was thought to correlate with the removal of the trigger. With enzymes that have the trigger residue orientated on a different subunit or domain to the immobile Walker A and B motifs, this could be seen as being coordinated with the movement of the piston; however, with enzymes that have the trigger residue located on the same subunit as the Walker A and B motifs, it would presumably be more difficult for the ADP molecule to be released. This is the case with many of the AAA⁺ family,[607,608] and as a result of this conformation, release of ADP molecules is difficult or even impossible without the interaction of other factors with the lid domain, which produce movement of this trigger residue away from the active site.

The final component of the different active sites was denoted as the tether, which was proposed to act much as a hinge between the two sides of the active site. Two forms of tether were identified: a stronger salt-bridge tether, comprising an acidic residue and the third basic residue, and a weaker glutamine-tethered active site. The glutamine tether comprises a conserved glutamine residue and an acid residue or alternatively was mediated through an aromatic residue and a conserved histidine, or an additional glutamine residue as was observed for the SF2 and SF4 family members.[598]

The different tethers appear to impose different structural limitations on the active site state and crystal structures of enzymes with the acid–base tether have been observed in all three possible conformations, whereas glutamine-tethered sites appear to have inherently more malleability, with the ATP-binding state almost indistinguishable. The best of these to have been structurally characterised is the SF4 T7gp4 helicase where the difference between ATP bound and empty sites is more apparent.[569]

4.4.10 Characterisation of Helicases/Translocases

Recent single-molecule studies with helicases have greatly improved our understandings of the motor activity of the enzymes.[609] Helicases and translocases, due to their often diverse functions, can be biochemically characterised based on a number of activities:

4.4.10.1 Rate of translocation

The rate of translocation can vary significantly from *in vitro* experiments based upon the nature of the substrate, the requirement for accessory factors, and can also be dependent on post-translational modification of the enzyme itself, and the rate of translocation is often directly linked to the ATPase activity of the enzyme and, therefore, step size (see below). *In vivo* many helicases have specific protein partners and target substrates that regulate their role in the cell.

4.4.10.2 Direction of translocation

Helicases have a specific direction of translocation along DNA, whether ssDNA or dsDNA. For the dsDNA, translocating enzymes experiments have shown that they tend to track along one specific strand in one direction whilst maintaining weak interactions with the complementary strand. Loading of the helicase onto the strand by partner proteins is thought to help orientate the helicase and, therefore, determine its direction of travel.

4.4.10.3 Processivity

Most helicases are thought to be in some way processive, i.e., they catalyse multiple reactions before releasing and rebinding their substrate DNA. As with the rate and direction of translocation, this characteristic is most likely also linked to other factors and accessory proteins that can modulate the enzyme.

4.4.10.4 Translocation step size

The distance of DNA translocated during each round of catalysis is often assumed to be directly proportional to ATP turnover; however, the exact step size relative to ATP hydrolysis may be more difficult to determine as the substrate can vary significantly in both base composition and structure. In addition, less frequent "kinetic" steps may occur that can skew these observations. Again, interaction with specific protein partners could also potentially give rise to different step sizes. However, motors such as helicases and the ribosome have been shown to involve 3-bp steps with three hidden sub-steps,[610] suggesting single-base steps for each ATP molecule consumed. Other motors, not discussed in this book, such as polymerases make specific single base-pair steps because they have to read the base-by-base content of their template. However, as described earlier, translocation by DNA-packaging motors (see earlier chapter) appears to involve the DNA helix (10-bp).

4.4.10.5 Active and passive enzymes

Although all helicases require ATP to function, the way in which the enzyme unwind DNA strands can be described as either active or passive.[611] Active helicases appear to actively participate in unwinding of the strands, whereas passive enzymes appear to take an opportunistic approach to unwinding, waiting for thermal fraying of the strands to occur before trapping the strands in their unwound state and further extending the fork.

4.4.11 Potential for Use of Biological Molecular Motors in Bionanotechnology

It seems clear from the literature that kinesins are believed to provide the greatest opportunity for future use of motors in bionanotechnology, and this can be summarised in tabular form as below.

Myosins are very diverse in their biological function, and many are very highly specialised, which suggests their general use in bionanotechnology may be somewhat limited. However, a clear capability is observed for a situation where a surface is coated with myosin and then used for actin fibre transport and manipulation[612] (c.f. kinesin and microtubules).

Dynein motors are also highly specialised, which again suggests their general use in bionanotechnology may be somewhat limited. However, these proteins are very large and often difficult to purify, while some structure/mechanisms are still to be elucidated.

Helicases are very important biomolecules with a wide range of functions that are closely related to DNA and RNA replication,[613] which has made them potential targets for drugs related to cancer.

Table 4.4 Summary of available biological molecular motor proteins

Name of motor	Enzyme/ gene name	Primary function	Source of protein	Most likely use in bionano-technology	Reference
Kinesins:					
Kinesin-1	KHC	Movement along the microtubule Positive-end directed	Several including Prof Robert Cross, University of Warwick.	Transport of cargo through movement of microtubules	614, 615, 616, 617, 306, 618,

(*Continued*)

Table 4.4 (*Continued*)

Name of motor	Enzyme/ gene name	Primary function	Source of protein	Most likely use in bionano- technology	Reference
Kinesin-14	KIF	Movement along the microtubule Negative-end directed	As above	Potential of variation in step size and speed	293, 446
Myosins:					
Myosin- II	Chicken Myosin-II Subfragment-1(motor domain)	Movement to positive end of actin fibre	Ivan Rayment, Department of Biochemistry and Institute for Enzyme Research, University of Wisconsin	Contraction and density changes through force production within materials	619
				Movement of actin fibres carrying cargo.	612
Non-muscle Myosin- II	Human Myosin-II	Movement to positive end of actin fibre.	Timothy J. Mitchison, Department of Systems Biology, Harvard Medical School, Boston	Movement of actin fibres carrying cargo	620
Myosin- VI	*Drosophila melanogaster* Myosin-VI	Negative-end-directed motion		As above	378, 379, 621
Myosin- IX	Human Myosin-IXb	Rapid motion of actin fibre	Mitsuo Ikebe, Department of Physiology, University of Massachusetts Medical School	Quicker motion in contractile materials	404, 408, 409

Name of motor	Enzyme/ gene name	Primary function	Source of protein	Most likely use in bionano-technology	Reference
Myosin- XI	Plant Myosin-XI	As above	Kazuhiro Oiwa, Kansai Advanced Research Center, Communications Research Laboratory, Kobe	As above	404, 408, 409
Dyneins:					
Cytoplasmic Dynein	Yeast Cytoplasmic Dynein	Negative-end-directed motility	Ronald D. Vale, Howard Hughes Medical Institute and Department of Cellular and Molecular Pharmacology, University of California	Transport of cargo through movement on/ of microtubules	483
Helicases and Translocases:					
SF1	*E. coli* UvrD	Unwinding nucleic acids	Timothy M Lohman, Department of Biochemistry and Molecular Biophysics, Washington University School of Medicine	Potential targets for cancer treatment	499, 622, 623, 624
Rep (SF1A)	Rep	As above	As above		502
EcoR124I	R.EcoR124I	DNA translocation /cleavage	Authors of this book	Single molecule nanoactuator	2

Chapter 5

Towards Bionanotechnology

There is little doubt that molecular motors provide a rich possibility for the development of nanoscale devices that are capable of useful work. K.F.'s work in this area started in the early 1970 when he began to study Type I restriction-modification (R-M) systems, or more specifically the EcoR124I Type IC R-M enzyme.[2] These complex enzymes were seen as interesting systems for "blue-sky research", but the fact that they were large, multi-subunit, multifunctional enzymes that cut DNA in a random and highly ATP-dependent manner suggested that they would not be a useful commercial enterprise (unlike the Type II restriction enzymes that led to the revolution in studies of biology that genetic engineering represented). However, as recently detailed,[2] the fact that the EcoR124I molecular motor was able to move micron-sized objects over several microns distance provided strong evidence that these motors could act as nanoactuators.[22,625] Even so, the idea to commercialise the use of a molecular motor as an active component of a biosensor has taken a number of years and a great deal of investment of time and funds and presents an interesting example of how to move from "blue skies" research to a commercial product. A similar situation has been obvious for a number of other biological molecular motors, where much of the research was carried out by cell biologists who were driven by a need to understand how many

Molecular Motors in Bionanotechnology
James Youell and Keith Firman

ISBN 978-981-4267-02-1 (Hardcover), 978-981-4303-09-5 (eBook)
www.panstanford.com

of the more complex tasks that a cell undertakes during its life cycle were accomplished, and which systems provided the driving force.

Once the level of understanding for these biological molecular motors reached the level described earlier in this book it became apparent that many of these systems might have commercial potential. Hess *et al.*[626] provided an early description of how biological molecular motor might be used in functional nanoscience and he imagined three main areas:

1. **Material transport** — where problems associated with nanofluidics might be overcome by using an active transport system in a manner analogous to the transport systems of the cell.
2. **Adaptive materials** — where biological molecular motors could be used to rearrange the structure of a material in a controllable manner. We would also include artificial muscle in such a definition and as indicated below, there is a great deal of potential in this area.
3. **Directed molecular assembly** — where a shuttle system is used as an artificial ribosome or molecular assembler[627,628] to provide directed arrangement of specific "building blocks".

The power and adaptability of biological molecular motors suggest that each of the above is possible, but before discussing applications of biological motors, it is of interest to first discuss the use of chemical motors in nanoscale devices.

5.1 Chemical Motors in Bionanotechnology

Perhaps one of the most ground-breaking applications of chemical motors was the use of rotaxanes, sandwiched between titanium and aluminium oxide electrodes as logic OR/AND arrays.[40,629,630] These systems worked through application of an oxidising voltage, which would change the oxidation (charge) state of one of the "docking" sites of the rotaxane molecule (see Chapter 2) and as a consequence a change to the conductance state of the molecular system. Gradual improvement to this simple system led to their development as molecular switch tunnel junctions[631] and by modification to the chemical design of the rotaxanes self-assembly of the devices was possible[632] with improved reliability of the devices[67] to produce a 64 bit random access memory device. There is no doubt that the use of such devices in future logic machines and computers represents an important breakthrough in the application of chemical motors, but wiring problems will undoubtedly require reliable application of connections at the nanoscale using carbon nanotubes and nanoscale wires.[633–635]

The previously described devices are really molecular machines that act as electronic devices, while much of our previous discussion about molecular motors has focussed on physical, directed motion. However, a device focussed on directed motion has been described in the form of a molecular elevator.[66] Such a device is in fact a nanoactuator based around the previously described shuttle systems that use rotaxanes[38] as the motor. The elevator device is 3.5 × 2.5 nm and contains a platform section (Fig. 5.1) that travels through 0.7 nm between the two levels that are defined by the chemical

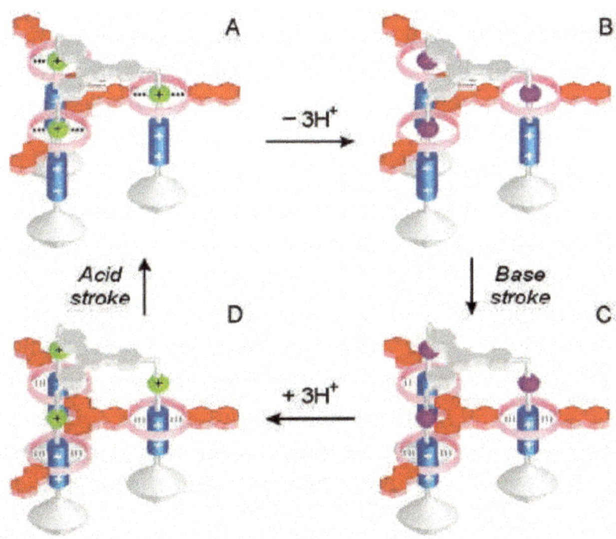

Figure 5.1 A molecular elevator. The molecular elevator consists of a tris-crown ether and a trifurcated guest salt $[4H_3][PF_6]_6$, which forms a 1:1 "superbundle" that can be converted into the mechanically interlocked elevator $[5H_3]_9$ by functionalisation of the ends of each leg with bulky 3,5-di-tert-butylbenzyl feet. The elevator is operated using a base/acid exchange system. Reproduced from http://stoddart.northwestern.edu/Index.php?ID=8&View=Research/ResearchSheet.php with permission from Dr Douglas C. Friedman.

state produced by an acid–base reaction, which also provides the energy for the motor. The force that this motor could generate has been estimated at 200 pN, which is potentially a much greater force than any of the previously described biological molecular motors, including DNA packaging motors (~70 pN) and bacterial flagella motors (~30 pN). The base of the elevator structure is rigid and chemically unavailable when the platform is at the lower position, but when the platform moves to the upper level, the base units are movable and chemically accessible. In addition, the platform can be linked covalently to other moieties suggesting a potential to manipulate objects using this device. The distance moved by the platform is very similar to the internal movement of myosin, which turn this small-scale movement into a 10 nm movement through a lever arm. Such a lever arm arrangement could potentially enable this device to readily manipulate microscale objects over a few tens of nanometres. That many such shuttle devices can be arrayed and consequently manipulate objects on the macroscale has already been clearly demonstrated[636] and their potential as artificial muscles is clear.

In summary, it is unlikely that chemical motors will generate useful devices for either material transport or directed molecular assembly (although they could provide a very localised means for positioning molecules for chemical reactions) due to the very short distance range over which such motors operate. Chemical motors can generate by far and away the greatest force per molecule and the fact that they can use important processes such as redox chemistry, electrochemistry, photoactivation, and pH as means of activation suggests they will have a large role to play in single-molecule biosensing. In addition, the fact that they can be arrayed in such a way that they can manipulate on the macroscale, or generate electronic devices, suggests that the short-range distances over which they can move objects are not a problem, and it is likely that power stroke devices will develop using a lever arm to amplify this movement. Arrayed devices would be a simple mechanism to interface such molecular transducers with macroscale devices that output useful signals (e.g., piezoelectric devices, or miniaturised cantilever systems). Therefore, we expect to see a significant number of chemical motors acting as transducers within biosensing system in the near future.

5.2 Biological Motors in Bionanotechnology

The wide range of biological motors that are either available, or have been studied to date, suggests that contributions towards all three suggested areas of application should be possible.[637] Soong *et al.*[638] described some of the problems associated with assembling multiple nanodevices, based around the use of ATP synthase, and how to overcome such problems by assembly of motors on "posts" etched into the surface. Therefore, we will discuss both theoretical and actual concepts for applying biological motors in the proposed areas:

Material Transport[639]: The problem for transporting materials on the nanoscale, identified by Hess *et al.*,[626] relates to the known difficulties in manipulating fluids within nanoscale channels and construction/metrology of such channels.[640,641] The requirement for high pressure pumping systems for channels less than 100 nm — such a channel requires a pressure of 1 Pa μm^{-1} for a flow rate of 1 $\mu m\ s^{-1}$ [642] — and the size of these pumps makes miniaturisation very difficult. Transport of both molecules and organelles, within a eukaryotic cell, makes use of microtubule (MT) structures[643] and the molecular motor kinesin to transport cargos along the MTs. However, it is difficult to grow MTs in such an organised way *in vitro* and researchers have shown that it is much easier to move the MT along a surface coated with the kinesin motors[644] (Fig. 5.2).

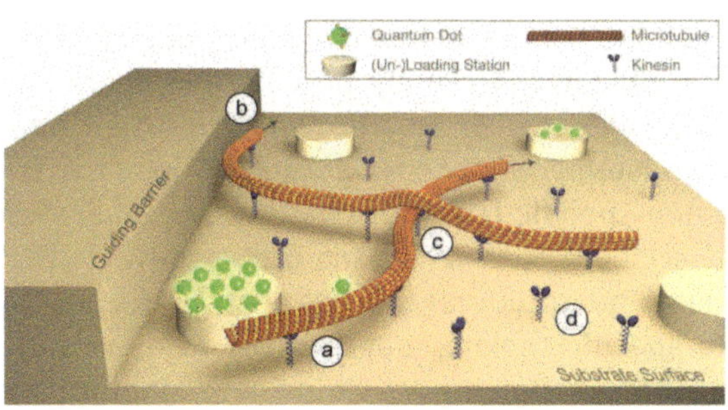

Figure 5.2 A microtubule-based transport system. A simple illustration of a cross-over event (c) for two microtubules moving across a surface coated with kinesin motors (d). In the proposed device, the microtubule would "pick-up" a cargo (quantum dot) from one station (a) and then transport this to an end station following a relatively direct route. Reproduced from bibliography reference[614] with permission from John Wiley and Sons.

The surface attached motors (e.g. kinesin-1) can transport the MT at ~1 μm s^{-1} and the system has been demonstrated for transport of a wide range of cargos from microspheres, quantum dots, DNA molecules to viruses[645–649] usually using a biotin-streptavidin linkage. Generally, movement of the MT is fairly straight as this is guided by the persistence length of the MT (which is ~100 μm), and the MT is held by the motors at a nearly constant height of 17 ± 2 nm. Of particular interest within a useful transport system is passage of one MT over/under another MT and this occurs without problems, but the mechanism (cross-over or cross-under) is dependent upon the degree of freedom of the advancing tip of the MT and at ~80 kinesin molecules μm^{-2}, almost 100% of such events are cross-overs. The MT movement can be guided along specific routes through surface etching and/or chemical modification of the surfaces[644]; however, the most obvious problem with such a device is programmed cargo release. This is exacerbated by the tight binding of the biotin-streptavidin linkage system that is often used. However, one can imagine the use of a chemical cross-linker that could be both photo-activated and also photo-cleavable using highly focussed laser light to enable such release at precise locations. Temporal control is possible using slowly hydrolysable analogues of ATP,[650] or through the use of thermo-responsive polymers.[651] In fact, the binding of some molecules ("cargo") has been observed to slow the rate of movement of the MT, suggesting a mechanism for using active transport to also sort materials.[652]

The advantage of the MT active-transport system illustrated in Fig. 5.2 is that it can readily be used to separate loaded and unloaded molecules, which may provide important improvements to biosensing systems, where special separation of a detection event can reduce non-specific false positives.[652] One such system, which has been described[616] as a variant of a sandwich ELISA, where antibody-coated microtubules that are centrally located on a surface divided into specific regions, but are free to move into the other areas of the surface following activation of MT-motility by activation of the surface-coated motors using caged ATP. In one such area, fluorescently-tagged secondary antibodies are localised and passage of an MT through this area will allow binding of the secondary antibody, which can then be detected by fluorescence output in the detection area. This system separates event spatially overcoming many of the problems associated with non-specific binding by the secondary antibodies (Fig. 5.3).

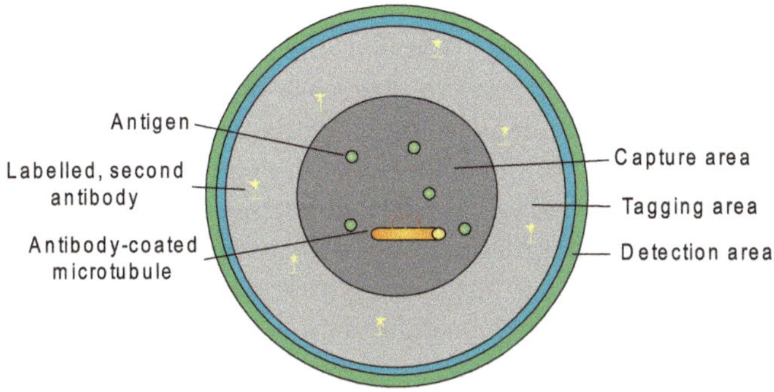

Figure 5.3 An active-transport sandwich ELISA system. A smart dust-detection system in which microtubules (MT) coated with an appropriate antibody can move into random areas of the device through an active-transport system of surface-coated kinesin-1 molecules. Fluorescently-labelled secondary antibodies, localised in the tagging area, will sandwich the antigen bound to the MT-antibody and will be detected in the detection area following excitation of the fluorophore. Adapted from bibliography reference.[652]

Other systems for transporting cargos can be imagined, and interest with DNA translocating motors has suggested their potential use to transport material attached to DNA. Such motors have been shown to be very processive,[625] and the ability to skip past gaps and be directed through DNA junctions is well documented.[653] Therefore, one can imagine precise localisation of branched DNA molecules using DNA-origami techniques[654] and the use of such motors to pull a dsDNA molecule (perhaps branched off a DNA-origami structure) and DNA-attached objects towards the motor (and consequently the surface where the origami DNA is attached). Of particular interest is the knowledge that these motors dislodge many proteins and cargos that are attached to the DNA, including Lac repressor molecules[487] and triple-helix forming oligonucleotides,[655] allowing a simple mechanism for delivery of the cargo. As discussed later in this chapter, such motors are manipulating a template that can carry information and the potential for use within an artificial ribosome becomes immediately apparent.

Copying some of nature's systems for organised transport is a sensible approach, but not the only means for creating an artificial transport system. Su *et al.*[202] described a unique, light-driven machine that also uses the molecular motor, ATP synthase, as a propeller for powered motion (Fig. 5.4). This machine perhaps represents one of the best examples of how to make use of nature's machines in a novel manner. Movement of the machine was observed through fluorescence tagging of the actin fibre, and speeds of a few tens of micrometres per second were observed, with an apparent swimming motion through cooperative activity of the assembled ATP synthase motors. It would be very interesting to know if this machine can swim "along" a laser light beam and as a consequence provide a mechanism for directed motion on the microscale. However, it is clear that the system is capable of transporting materials attached to the actin fibre. Such a machine also points towards future devices for synthetic biology, where light could be converted into ATP using biological motors and thence into electricity through movement of magnetic beads,[2] or manipulation of piezoelectrical devices and electroactive polymers.

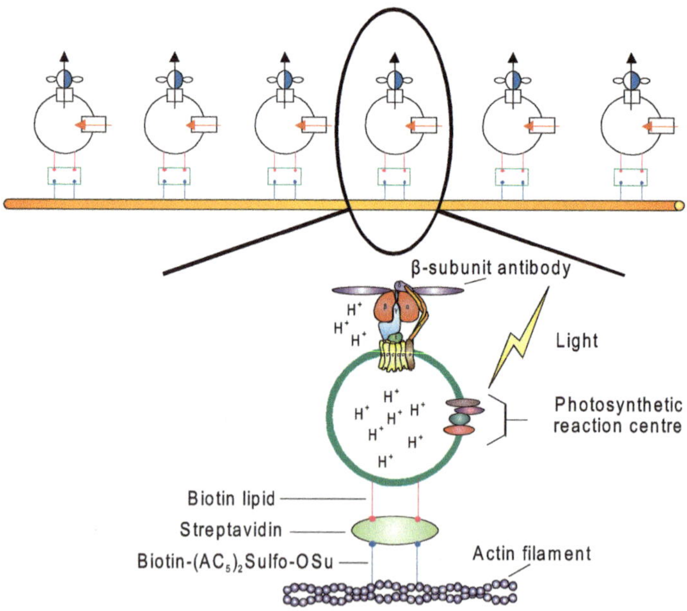

Figure 5.4 A light-driven machine. A nanodevice constructed from ATP synthase motors assembled in series along an actin fibre, which is able to use light to produce physical work in the form of a swimming motion produced through a propeller-like arrangement of the motors.[202] Adapted from reference [202].

Our own work can also be mentioned in this area as we have shown that the Type IC restriction-modification enzyme EcoR124I can be used as a unidirectional molecular motor that can "pull" dsDNA over long distances (several micrometres) even when the end of the DNA is attached to a micron-sized paramagnetic bead. We have described such a device as a molecular dynamo because the moving magnetic bead can be detected using appropriate sensors. In fact the activity of the motor has been characterised using an optical system (a magnetic tweezer set-up)[22,23,625]; but our work with a European research group has shown that an electronic version of such a system can be constructed,[656] and we have suggested that such a device could be used in the area of drug discovery, because many DNA manipulating enzymes are also targets for drugs (e.g. topoisomerase I[657]). An exciting area for such a device would be re-screening existing drugs as well as new compounds against chromatin remodelling complexes, which have been recognised as potential targets for anticancer therapy,[658] or within more complex systems such as histone acetylation assays.[659] The proposed device offers two important opportunities in the area of drug discovery: rapid screening of drugs at the single-molecule level (where sensitivity is based on the kd of the drug–target system) and electronic output from the biological sensing system at the single-molecule level. It is clear that the application of single molecule techniques, within the area of biosensing and biomedical devices, will significantly improve sensing capabilities and will provide a paradigm shift in levels of sensitivity.

Adaptive Materials: A major goal of nanotechnology is to develop an intelligent material that responds to local environmental change, especially involving the human body, providing both support and protection in situations ranging from stress to injury. Such materials are likely to involve "*shape-change*" and density changes, which, in turn, would suggest uses in other environments (a good example is targeted drug release). While pH and thermally responsive polymers have been described,[660–665] truly intelligent materials will require additional responsive elements to enable changes to the characteristics of the material over a wide range of conditions. A possible route to the production of such novel materials is through the incorporation of molecular motors[666] that can manipulate fibres. To date we are not aware of any successful demonstrations of such a material; although, an example of a development in this general direction is the use of biomolecular linkers to control the responsive nature of the material, where the cross-linking agent was a peptide.[667] However, DNA seems to offer a greater number of possibilities for both nanostructuring and design.[668,669] As detailed above, we have considerable experience of using DNA translocating motors within commercial devices, and we also imagine a variation of this work where DNA-cross-linked polymers could be manipulated, to change their density or packaging orientation, through translocation of the cross-linking DNA molecules. These motors respond to ATP, which is available on the skin surface and such a material can be imagined as providing useful tourniquets and support materials.

Directed Molecular Assembly: The concept of molecular assembly was first detailed by Richard Feynman in his famous talk "*plenty of room at the bottom*"[670] and the idea further developed by Eric Drexler,[627,671] where he described a machine that might mimic the ribosome in the ability to assemble materials, but be independent of the requirement for only a limited type of building block. There are several problems that must be solved before such a device can be produced, but a number of such steps have now been made possible. Perhaps the first main step, that would be necessary for such a device to function efficiently, is that it must be possible to carry out a very localised chemistry, which would allow different building blocks to be joined together. Recent progress has been made in providing a variety of techniques for single-molecule manipulations, including the use of DNA origami to provide localised chemistry,[50] the use of dip pen nanolithography (DPN) to deliver specific molecules to a limited location[672] and precise positioning using the AFM.[673] It is now possible to imagine a simple conveyor-belt artificial ribosome as mentioned by Hess *et al.*,[626] and perhaps the use of DNA as the template of instructions that guide the order of delivery of building blocks. In such a system, a motor that "pulls" DNA past a variety of gaps and junctions would be useful[653] as this would allow use of a ssDNA template with attached, complementary oligonucleotides carrying the building blocks guaranteed (through the DNA sequence of the template) to deliver the building blocks in the correct order. Displacement of the oligonucleotide[655] would release the building blocks for a highly localised chemistry and the machine could be controlled by the use of caged ATP.[648,674] However, development of any such system remains a distant dream at this time.

5.3 Bionanotechnology Devices *in vivo*

One of the major challenges in developing moving devices for bionanotechnology is to demonstrate their use and applicability *in vivo*, and the DNA-based device described by Modi *et al.*,[56] described briefly in Chapter 2, is one of the few nanodevices that have been shown to operate in a predictable manner inside a cell. In effect, this device is a miniature pH probe that operates over the physiologically important pH range 5.5–6.8 and, specifically, signals the presence of protons through a fluorescent output over an output time of 1–2 min. This device was demonstrated *in vivo* by following the characteristic pH changes associated with endosome maturation[675] in *Drosophila* haemocytes, which appear in the confocal microscope as fluorescent output within the cell (Fig. 5.5). However, the system was also used to localise specific proteins during pH changes, and potentially protein pathways of interest, through biotinylation of the protein of interest and biotinylation of the DNA machine, which could then be coupled to each other through a streptavidin sandwich.

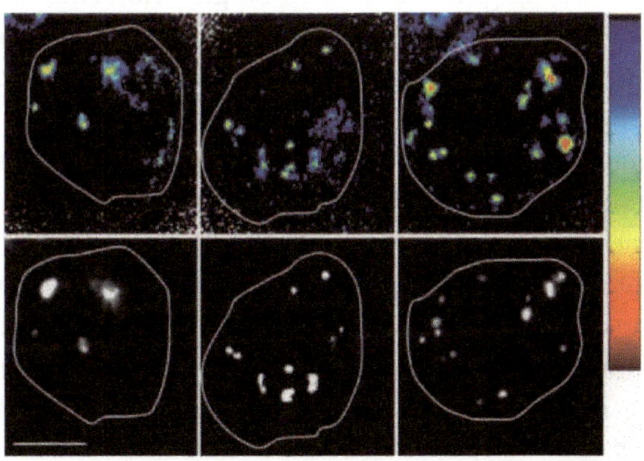

Figure 5.5 Cellular localisation of a DNA nanomachine. The fluorescent output from the FRET signals of the DNA machine show both spatial and temporal changes in pH during endocytosis of living cells. Reprinted by permission from Macmillan Publishers Ltd from reference [56]. Copyright 2009.

This type of device suggests a huge potential in biomedical uses and especially monitoring abnormal gene expression and localised environmental changes within a living cell. Undoubtedly, this points the way for the future use of molecular machines.

5.4 Conclusion

Significant progress has already been made in the use of molecular motors in electronics and for directed motion/transport of materials, while it is now already possible to readily imagine more exotic uses ranging from responsive materials to a molecular assembler. The guiding influence in many projects has been nature's own machinery, and this remains a sensible and important aspect of work in this area. It is of ultimate importance that current "blue-skies" research is allowed to develop our understanding of biological molecular motors as well as in the area of synthesis of chemical motors as this clearly leads to commercial opportunities and novel uses of the motors under study. Nanotechnology provides an important concept to the development of novel devices and that concept leads us back to our early use of a definition of nanotechnology — the use of single molecules — and perhaps this is where a huge commercial opportunity exists. The development of bionanotechnology has revolved around the ability to detect and measure single-molecule activity (optical tweezers, magnetic tweezers, single- molecule fluorescence and force microscopy), and this capability offers an opportunity to apply the techniques in biosensing. Many of the existing devices do a very effective job that is unlikely to be greatly improved by the use of novel technology unless such technology greatly improves sensitivity and, of course, a device that can detect events at the single-molecule level is already at the limits of sensitivity.

The use of motors to manipulate and position such single molecules, within novel devices, would be a very productive area of the use of kinesins, and myosins within this area looks very feasible and one can readily foresee active transport systems built around these motors.

Many DNA-manipulating enzymes are also molecular motors, but are also targets for anticancer and other drugs. There has been significant progress in the handling and measurement of motor activity of these enzymes, which suggests a route to development of single molecule biosensors for use in drug screening. Such devices would offer new opportunities for the detection of drug–target interactions at the very limit of sensitivity based on kd for such interactions.

Rotary motors, both chemical and biological, have the potential for the development of free-swimming devices, many of which would swim along a light gradient, providing a mechanism for guided movement of objects ranging from a few nanometres to several micrometres in size. One can also imagine such spinning devices as a means for generating electricity, by rotation of magnetic nanoparticles and there seems a likely growth in the conversion of light to mechanical rotation and subsequently the generation of electricity. Whether the search for alternative sources of power will make use of molecular motors remains to be seen and whether their efficiency would justify such investment remains uncertain at this time, but mimicking nature is seem as a useful approach to such technological developments.

Linear chemical motors have already been shown to be capable of acting on the macroscale in a manner that is reminiscent of artificial muscles, and it seems likely that other motors that "pull" molecules such as DNA (e.g. Type I restriction-modification enzymes) could also manipulate materials to which the DNA is attached. Therefore, a likely growth area for the use of molecular motors is in the development of intelligent materials, which react to environmental changes in a more controlled manner, with fewer limitations, than those currently possible based on the chemistry of the material.

However, it is also clear that not all motors have the potential for use in bionanotechnology and that many of the kinesins, dyneins, and myosins may only be the subject of study of cell biology. Yet, the greatly improved understanding of both structure and function of these motors, in part resulting from single-molecule work, may make them targets for drug design for certain diseases. The ability to measure motor activity at the single-molecule level will, once again, assist in the identification of suitable drugs and improve our understanding of how these drugs interfere with motor activity. The suggestion that existing drugs should be re-screened for new targets using new technologies[676] supports this concept and would appear to be a useful way to develop existing resources. It seems likely that such studies will eventually be seen as bionanotechnology in their own right.

Bibliography

1. Wang, M. D. (1999). Manipulation of single molecules in biology. *Current Opinions in Biotechnology* **10**, 81–86.
2. Youell, J., and Firman, K. (2008). EcoR124I: from plasmid-encoded restriction–modification system to nanodevice. *Microbiology and Molecular Biology Reviews* **72**, 365–377.
3. Kay, E. R., Leigh, D. A., and Zerbetto, F. (2007). Synthetic molecular motors and mechanical machines. *Angewandte Chemie International Edition* **46**, 72–191.
4. Singleton, M. R., Dillingham, M. S., and Wigley, D. B. (2007). Structure and mechanism of helicases and nucleic acid translocases. *Annual Review of Biochemistry* **76**, 23–50.
5. Banting, G., and Higgins, S. J. (2000). *Essays in Biochemistry: Molecular Motors*, **35**, Portland Press, London.
6. Schliwa, M. (ed). (2003). *Molecular Motors*, 1st ed, Wiley-VCH, Weinheim.
7. Oster, G., and Wang, H. (2003). How protein motors convert chemical energy into mechanical work, in *Molecular Motors* (Schliwa, M., ed), Wiley-VCH, Weinheim, pp. 207–227.
8. Astumian, R. D. (2000). The role of thermal activation in motion and force generation by molecular motors. *Philosophical Transactions of the Royal Society of London. Series B: Biological Sciences* **355**, 511–22.
9. Ait-Haddou, R., and Herzog, W. (2003). Brownian ratchet models of molecular motors. *Cell Biochemistry and Biophysics* **38**, 191–214.
10. Reimann, P. (2002). Brownian motors: noisy transport far from equilibrium. *Physics Reports* **361**, 57–265.
11. Chacon, R., and Quintero, N. R. (2007). Controlling the ratchet effect through the symmetries of the systems: Application to molecular motors. *Biosystems* **88**, 308–315.
12. Hoki, K., Yamaki, M., Koseki, S., and Fujimura, Y. (2003). Molecular motors driven by laser pulses: Role of molecular chirality and photon helicity. *Journal of Chemical Physics* **118**, 497–504.
13. Johansson, M. P., and Patzschke, M. (2009). Fixing the chirality and trapping the transition state of helicene with atomic metal glue. *Chemistry — A European Journal* **15**, 13210–13218.

14. Schalley, C. A., Beizai, K., and Vogtle, F. (2001). On the way to rotaxane-based molecular motors: Studies in molecular mobility and topological chirality. *Accounts of Chemical Research* **34**, 465–476.

15. Wang, H., and Oster, G. (2002). The Stokes efficiency for molecular motors and its applications. *Europhysics Letters* **57**, 134–140.

16. Fisher, M. E., and Kolomeisky, A. B. (1999). The force exerted by a molecular motor. *Proceedings of the National Academy of Sciences of the United States of America* **96**, 6597–6602.

17. Doering, C., Ermentrout, B., and Oster, G. (1995). Rotary DNA motors. *Biophysical Journal* **69**, 2256–2267.

18. Little, M. A., Steel, B. C., Bai, F., Sowa, Y., Bilyard, T., Mueller, David M., Berry, Richard M., and Jones, Nick S. (2011). Steps and bumps: Precision extraction of discrete states of molecular machines. *Biophysical Journal* **101**, 477–485.

19. Ballardini, R., Balzani, V., Credi, A., Gandolfi, M. T., and Venturi, M. (2001). Artificial molecular-level machines: Which energy to make them work? *Accounts of Chemical Research* **34**, 445–455.

20. Marx, B. (2001). Optical tweezers help study molecular motors. *Laser Focus World* **37**, 30–33.

21. Li, T., Kheifets, S., Medellin, D., and Raizen, M. G. (2010). Measurement of the instantaneous velocity of a Brownian particle. *Science* **328**, 1673–1675.

22. Seidel, R., van Noort, J., van der Scheer, C., Bloom, J. G. P., Dekker, N. H., Dutta, C. F., Blundell, A., Robinson, T., Firman, K., and Dekker, C. (2004). Real-time observation of DNA translocation by the Type I restriction-modification enzyme EcoR124I. *Nature Structural Molecular Biology* **11**, 838–843.

23. Seidel, R., Bloom, J. G., Dekker, C., and Szczelkun, M. D. (2008). Motor step size and ATP coupling efficiency of the dsDNA translocase EcoR124I. *The EMBO Journal* **27**, 1388–1398.

24. Wang, H., and Oster, G. (2002). Ratchets, power strokes, and molecular motors. *Applied Physics A: Materials Science and Processing* **75**, 315–324.

25. Kinosita, K. J., Yasuda, R., Noji, H., and Adachi, K. (2000). A rotary molecular motor that can work at near 100% efficiency. *Philosophical Transactions of the Royal Society of London. Series B: Biological Sciences* **355**, 473–89.

26. Xu, B., *et al.* (1999). Study on the viscosity of the liquid flowing in microgeometry. *Journal of Micromechanics and Microengineering* **9**, 377.

27. Hernandez, J. V., Kay, E. R., and Leigh, D. A. (2004). A reversible synthetic rotary molecular motor. *Science* **306**, 1532–1537.

28. Keller, D., and Bustamante, C. (2000). The mechanochemistry of molecular motors. *Biophysical Journal* **78**, 541–556.

29. Kelly, T. R., De Silva, H., and Silva, R. A. (1999). Unidirectional rotary motion in a molecular system. *Nature* **401**, 150–152.

30. Koumura, N., Zijlstra, R. W., van Delden, R. A., Harada, N., and Feringa, B. L. (1999). Light-driven monodirectional molecular rotor. *Nature* **401**, 152–155.

31. Kelly, T. R., Silva, R. A., De Silva, H., Jasmin, S., and Zhao, Y. J. (2000). A rationally designed prototype of a molecular motor. *Journal of the American Chemical Society* **122**, 6935–6949.

32. Kelly, T. R., Cai, X. L., Damkaci, F., Panicker, S. B., Tu, B., Bushell, S. M., Cornella, I., Piggott, M. J., Salives, R., Cavero, M., Zhao, Y. J., and Jasmin, S. (2007). Progress toward a rationally designed, chemically powered rotary molecular motor. *Journal of the American Chemical Society* **129**, 376–386.

33. Koumura, N., Geertsema, E. M., van Gelder, M. B., Meetsma, A., and Feringa, B. L. (2002). Second-generation light-driven molecular motors. Unidirectional rotation controlled by a single stereogenic center with near-perfect photoequilibria and acceleration of the speed of rotation by structural modification. *Journal of the American Chemical Society* **124**, 5037–5051.

34. van Delden, R. A., ter Wiel, M. K., Pollard, M. M., Vicario, J., Koumura, N., and Feringa, B. L. (2005). Unidirectional molecular motor on a gold surface. *Nature* **437**, 1337–1340.

35. ter Wiel, M. K., van Delden, R. A., Meetsma, A., and Feringa, B. L. (2003). Increased speed of rotation for the smallest light-driven molecular motor. *Journal of the American Chemical Society* **125**, 15076–15086.

36. Liu, Y., Flood, A. H., Bonvallett, P. A., Vignon, S. A., Northrop, B. H., Tseng, H. R., Jeppesen, J. O., Huang, T. J., Brough, B., Baller, M., Magonov, S., Solares, S. D., Goddard, W. A., Ho, C. M., and Stoddart, J. F. (2005). Linear artificial molecular muscles. *Journal of the American Chemical Society* **127**, 9745–9759.

37. Serreli, V., Lee, C.-F., Kay, E. R., and Leigh, D. A. (2007). A molecular information ratchet. *Nature* **445**, 523–527.

38. Anelli, P. L., Spencer, N., and Stoddart, J. F. (1991). A molecular shuttle. *Journal of the American Chemical Society* **113**, 5131–5133.

39. Huang, T. J., Brough, B., Ho, C.-M., Liu, Y., Flood, A. H., Bonvallet, P. A., Tseng, H.-R., and Stoddart, J. F. (2004). A nanomechanical device based on linear molecular motors. *Applied Physics Letters* **85**, 5391–5393.

40. Harada, A. (2001). Cyclodextrin-based molecular machines†. *Accounts of Chemical Research* **34**, 456–464.

41. Bath, J., and Turberfield, A. J. (2007). DNA nanomachines. *Nature Nanotechnology* **2**, 275–284.

42. Gu, H., Chao, J., Xiao, S.-J., and Seeman, N. C. (2009). Dynamic patterning programmed by DNA tiles captured on a DNA origami substrate. *Nature Nanotechnology* **4**, 245–248.

43. Jungmann, R., Liedl, T., Sobey, T. L., Shih, W., and Simmel, F. C. (2008). Isothermal assembly of DNA origami structures using denaturing agents. *Journal of the American Chemical Society* **130**, 10062–10063.

44. Seeman, N. C. (1993). Branched DNA: A 3-D structural design system. *Clinical Chemistry* **39**, 722.

45. Seeman, N. C. (1993). Nanoengineering with DNA. In *Biomolecular Materials* (Viney, C., Case, S. T., and Waite, J. H., eds), Vol. 292, p. 123. Materials Research Society.

46. Seeman, N. C., Y. Zhang & J. Chen, (1994). DNA nanoconstructions. *Journal of Vacuum Science and Technology a Vacuums Surfaces and Films* **12**, 1895.

47. Seeman, N. C., Zhang, Y., Shou Ming, D., and Hui, W. (1994). The control of DNA structure and topology: An overview. In *Molecularly Designed Ultrafine/Nanostructured Materials* (Gonsalves, K. E., Chow, G. M., Xiao, T. D., and Cammarata, R. C., eds), Vol. 351, p. 57. Materials Research Society.

48. Seeman, N. C. (2007). An overview of structural DNA nanotechnology. *Molecular Biotechnology* **37**, 246–257.

49. Xiao, S., Liu, F., Rosen, A. E., Hainfeld, J. F., Seeman, N. C., Musier-Forsyth, K., and Kiehl, R. A. (2002). Self-assembly of metallic nanoparticle arrays by DNA scaffolding. *Journal of Nanoparticle Research* **4**, 313–317.

50. Voigt, N. V., Torring, T., Rotaru, A., Jacobsen, M. F., Ravnsbaek, J. B., Subramani, R., Mamdouh, W., Kjems, J., Mokhir, A., Besenbacher, F., and Gothelf, K. V. (2010). Single-molecule chemical reactions on DNA origami. *Nature Nanotechnology* **5**, 200–203.

51. Chakraborty, B., Sha, R., and Seeman, N. C. (2008). A DNA-based nanomechanical device with three robust states. *Proceedings of the National Academy of Sciences of the United States of America* **105**, 17245–17249.

52. Mao, C., Sun, W., Shen, Z., and Seeman, N. C. (1999). A nanomechanical device based on the B-Z transition of DNA. *Nature* **397**, 144–146.

53. Yurke, B., Turberfield, A. J., Mills, A. P., Simmel, F. C., and Neumann, J. L. (2000). A DNA-fuelled molecular machine made of DNA. *Nature* **406**, 605–608.

54. Muller, B. K., Reuter, A., Simmel, F. C., and Lamb, D. C. (2006). Single-pair FRET characterization of DNA tweezers. *Nano Letters* **6**, 2814–2820.

55. Zhang, H., Yu, H., Ren, J., and Qu, X. (2006). Reversible B/Z-DNA transition under the low salt condition and non-B-form polydApolydT selectivity by a cubane-like europium-L-aspartic acid complex. *Biophysical Journal* **90**, 3203–3207.

56. Modi, S., Swetha, M. G., Goswami, D., Gupta, G. D., Mayor, S., and Krishnan, Y. (2009). A DNA nanomachine that maps spatial and temporal pH changes inside living cells. *Nature Nanotechnology* **4**, 325–330.

57. Ishitsuka, Y., and Ha, T. (2009). DNA nanotechnology: A nanomachine goes live. *Nature Nanotechnology* **4**, 281–282.

58. Turberfield, A. J., Mitchell, J. C., Yurke, B., Mills, A. P., Jr., Blakey, M. I., and Simmel, F. C. (2003). DNA fuel for free-running nanomachines. *Physical Review Letters* **90**, 118102.

59. Yin, P., Yan, H., Daniell, X. G., Turberfield, A. J., and Reif, J. H. (2004). A unidirectional DNA walker that moves autonomously along a track. *Angewandte Chemie International Edition in English* **43**, 4906–4911.

60. Yin, P., Turberfield, A. J., and Reif, J. H. (2005). Designs of autonomous unidirectional walking DNA devices. *Lecture Notes in Computer Science* **3384**, 410–425.

61. Yin, P., Turberfield, A. J., Sahu, S., and Reif, J. H. (2005). Design of an autonomous DNA nanomechanical device capable of universal computation and universal translational motion. *Lecture Notes in Computer Science* **3384**, 426–444.

62. Shin, J. S., and Pierce, N. A. (2004). A synthetic DNA walker for molecular transport. *Journal of the American Chemical Society* **126**, 10834–10835.

63. Sherman, W. B., and Seeman, N. C. (2004). A precisely controlled DNA biped walking device. *Nano Letters* **4**, 1203–1207.

64. Tian, Y., He, Y., Chen, Y., Yin, P., and Mao, C. D. (2005). Molecular devices: A DNAzyme that walks processively and autonomously along a one-dimensional track. *Angewandte Chemie International Edition in English* **44**, 4355–4358.

65. Bath, J., Green, S. J., and Turberfield, A. J. (2005). A free-running DNA motor powered by a nicking enzyme. *Angewandte Chemie International Edition in English* **44**, 4358–4361.

66. Badjic, J. D., Balzani, V., Credi, A., Silvi, S., and Stoddart, J. F. (2004). A molecular elevator. *Science* **303**, 1845–1849.

67. Luo, Y., Collier, C. P., Jeppesen, J. O., Nielsen, K. A., DeIonno, E., Ho, G., Perkins, J., Tseng, H.-R., Yamamoto, T., Stoddart, J. F., and Heath, J. R. (2002). Two-dimensional molecular electronics circuits. *Chemphyschem* **3**, 519–525.

68. Omabegho, T., Sha, R., and Seeman, N. C. (2009). A bipedal DNA Brownian motor with coordinated legs. *Science* **324**, 67–71.

69. Li, X., Yang, X., Qi, J., and Seeman, N. C. (1996). Antiparallel DNA double crossover molecules as components for nanoconstruction. *Journal of the American Chemical Society* **118**, 6131–6140.

70. von Ballmoos, C., Wiedenmann, A., and Dimroth, P. (2009). Essentials for ATP synthesis by F_1F_0-ATP synthases. *Annual Review of Biochemistry* **78**, 649–672.

71. Boyer, P. D., and Kohlbrenner, W. E. (1981). The present status of the binding-change mechanism and its relation to ATP formation by chloroplasts. In *Energy Coupling in Photosynthesis* (Selman, B. R., and Selman-Reimer, S., eds), pp. 231–240, Elsevier, Amsterdam.

72. Abrahams, J. P., Leslie, A. G. W., Lutter, R., and Walker, J. E. (1994). Structure at 2.8 Å of F_1-ATPase from bovine heart mitochondria. *Nature* **370**, 621–628.

73. Nakamoto, R. K., Baylis Scanlon, J. A., and Al-Shawi, M. K. (2008). The rotary mechanism of the ATP synthase. *Archives of Biochemistry and Biophysics* **476**, 43–50.

74. Welch, A. K., Bostwick, C. J., and Cain, B. D. (2011). Manipulations in the peripheral stalk of the saccharomyces cerevisiae F_1F_0-ATP synthase. *The Journal of Biological Chemistry* **286**, 10155–10162.

75. Kohori, A., Chiwata, R., Hossain, Mohammad D., Furuike, S., Shiroguchi, K., Adachi, K., Yoshida, M., and Kinosita Jr., K. (2011). Torque generation in F_1-ATPase devoid of the entire amino-terminal helix of the rotor that fills half of the stator orifice. *Biophysical Journal* **101**, 188–195.

76. Wächter, A., Bi, Y., Dunn, S. D., Cain, B. D., Sielaff, H., Wintermann, F., Engelbrecht, S., and Junge, W. (2011). Two rotary motors in F-ATP synthase are elastically coupled by a flexible rotor and a stiff stator stalk. *Proceedings of the National Academy of Sciences of the United States of America* **108**, 3924–3929.

77. Kim, Y., Konno, H., Sugano, Y., and Hisabori, T. (2011). Redox regulation of rotation of the cyanobacterial F_1-ATPase containing thiol regulation switch. *The Journal of Biological Chemistry* **286**, 9071–9078.

78. Kobayashi, M., Akutsu, H., Suzuki, T., Yoshida, M., and Yagi, H. (2010). Analysis of the open and closed conformations of the [beta] subunits in

thermophilic F_1-ATPase by solution NMR. *Journal of Molecular Biology* **398**, 189–199.

79. Yasuda, R., Noji, H., Yoshida, M., Kinosita, K., and Itoh, H. (2001). Resolution of distinct rotational substeps by submillisecond kinetic analysis of F_1-ATPase. *Nature* **410**, 898–904.

80. Ishmukhametov, R., Hornung, T., Spetzler, D., and Frasch, W. D. (2010). Direct observation of stepped proteolipid ring rotation in *E. coli* F_0F_1-ATP synthase. *The EMBO Journal* **29**, 3911–3923.

81. Shimabukuro, K., Yasuda, R., Muneyuki, E., Hara, K. Y., Kinosita, K., Jr., and Yoshida, M. (2003). Catalysis and rotation of F_1 motor: Cleavage of ATP at the catalytic site occurs in 1 ms before 40 degree substep rotation. *Proceedings of the National Academy of Sciences of the United States of America* **100**, 14731–14736.

82. Nakamoto, R. K., Ketchum, C. J., and al-Shawi, M. K. (1999). Rotational coupling in the F_0F_1-ATP synthase. *Annual Review of Biophysics and Biomolecular Structure* **28**, 205–234.

83. Kaibara, C., Matsui, T., Hisabori, T., and Yoshida, M. (1996). Structural asymmetry of F_1-ATPase caused by the gamma subunit generates a high affinity nucleotide binding site. *The Journal of Biological Chemistry* **271**, 2433–2438.

84. Feniouk, B. A., Kato-Yamada, Y., Yoshida, M., and Suzuki, T. (2010). Conformational transitions of subunit in ATP synthase from thermophilic bacillus PS3. *Biophysical Journal* **98**, 434–442.

85. Shimo-Kon, R., Muneyuki, E., Sakai, H., Adachi, K., Yoshida, M., and Kinosita, K. (2010). Chemo-mechanical coupling in F_1-ATPase revealed by catalytic site occupancy during catalysis. *Biophysical Journal* **98**, 1227–1236.

86. Weber, J., Bowman, C., and Senior, A. E. (1996). Specific tryptophan substitution in catalytic sites of *Escherichia coli* F_1-ATPase allows differentiation between bound substrate ATP and product ADP in steady-state catalysis. *The Journal of Biological Chemistry* **271**, 18711–18718.

87. Beke-Somfai, T., Lincoln, P., and Nordén, B. (2011). Double-lock ratchet mechanism revealing the role of αSER-344 in F_0F_1-ATP synthase. *Proceedings of the National Academy of Sciences of the United States of America* **108**, 4828–4833.

88. Itoh, H., Takahashi, A., Adachi, K., Noji, H., Yasuda, R., Yoshida, M., and Kinosita, K. (2004). Mechanically driven ATP synthesis by F_1-ATPase. *Nature* **427**, 465–468.

89. Perez-Carrasco, R., and Sancho, J. M. (2010). Theoretical analysis of the F_1-ATPase experimental data. *Biophysical Journal* **98**, 2591–2600.

90. Feniouk, B. A., Suzuki, T., and Yoshida, M. (2006). The role of subunit epsilon in the catalysis and regulation of F_0F_1-ATP synthase. *Biochimica et Biophysica Acta (BBA)-Bioenergetics 14th European Bioenergetics Conference 2006 Short Reports* **1757**, 326–338.

91. Richter, M. L. (2004). Gamma-epsilon interactions regulate the chloroplast ATP synthase. *Photosynthesis Research* **79**, 319–329.

92. Cingolani, G., and Duncan, T. M. (2011). Structure of the ATP synthase catalytic complex (F_1) from *Escherichia coli* in an autoinhibited conformation. *Nature Structural and Molecular Biology* **18**, 701–707.

93. Noji, H., Yasuda, R., Yoshida, M., and Kinosita Jr., K. (1997). Direct observation of the rotation of F_1-ATPase. *Nature* **386**, 299–302.

94. Kinosita Jr., K. (1998). Linear and rotary molecular motors. In *Mechanisms of Work Production and Work Absorption in Muscle* (Sugi, H., and Pollack, G. H., eds), Vol. 453, Plenum Press, New York and London, pp. 5–14.

95. Kinosita Jr., K., Yasuda, R., and Noji, H. (2000). F_1-ATPase: A highly efficient rotary ATP machine. In *Essays in Biochemistry: Molecular Motors* (Banting, G., and Higgins, S. J., eds), Portland Press, London, pp. 3–18.

96. Girvin, M. E., Rastogi, V. K., Abildgaard, F., Markley, J. L., and Fillingame, R. H. (1998). Solution structure of the transmembrane H^+-transporting subunit c of the F_1F_0-ATP synthase. *Biochemistry* **37**, 8817–8824.

97. Stock, D., Gibbons, C., Arechaga, I., Leslie, A. G. W., and Walker, J. E. (2000). The rotary mechanism of ATP synthase. *Current Opinion in Structural Biology* **10**, 672–679.

98. Dimroth, P., von Ballmoos, C., Meier, T., and Kaim, G. (2003). Electrical power fuels rotary ATP synthase. *Structure* **11**, 1469–1473.

99. Rastogi, V. K., and Girvin, M. E. (1999). Structural changes linked to proton translocation by subunit c of the ATP synthase. *Nature* **402**, 263–268.

100. Dong, H., and Fillingame, R. H. (2010). Chemical reactivities of cysteine substitutions in subunit a of ATP synthase define residues gating H^+ Transport from each side of the membrane. *The Journal of Biological Chemistry* **285**, 39811.

101. Cain, B. D., and Simoni, R. D. (1986). Impaired proton conductivity resulting from mutations in the a subunit of F_1F_0-ATPase in *Escherichia coli*. *The Journal of Biological Chemistry* **261**, 10043–10050.

102. Sambongi, Y., Iko, Y., Tanabe, M., Omote, H., Iwamoto-Kihara, A., Ueda, I., Yanagida, T., Wada, Y., and Futai, M. (1999). Mechanical rotation of the c subunit oligomer in ATP synthase (F_0F_1): direct observation. *Science* **286**, 1722–1724.

103. Panke, O., Gumbiowski, K., Junge, W., and Engelbrecht, S. (2000). F-ATPase: specific observation of the rotating c subunit oligomer of EF(0)EF(1). *FEBS Letters* **472**, 34–38.

104. Nishio, K., Iwamoto-Kihara, A., Yamamoto, A., Wada, Y., and Futai, M. (2002). Subunit rotation of ATP synthase embedded in membranes: A or beta subunit rotation relative to the c subunit ring. *Proceedings of the National Academy of Sciences of the United States of America* **99**, 13448–13452.

105. Zimmermann, B., Diez, M., Borsch, M., and Graber, P. (2006). Subunit movements in membrane-integrated EF_0F_1 during ATP synthesis detected by single-molecule spectroscopy. *Biochimica et Biophysica Acta (BBA) — Bioenergetics 14th European Bioenergetics Conference 2006 Short Reports* **1757**, 311–319.

106. Börsch, M. (2010). Single-molecule fluorescence resonance energy transfer techniques on rotary ATP synthases. *Biological Chemistry* **392**, 135–142.

107. Aldridge, C., Poonchareon, K., Saini, S., Ewen, T., Soloyva, A., Rao, C. V., Imada, K., Minamino, T., and Aldridge, P. D. (2010). The interaction dynamics of a negative feedback loop regulates flagellar number in Salmonella enterica serovar Typhimurium. *Molecular Microbiology*, **78**(6), 1416–1430.

108. Ling, H., Kang, A., Tan, M. H., Qi, X., and Chang, M. W. (2010). The absence of the luxS gene increases swimming motility and flagella synthesis in *Escherichia coli* K12. *Biochemical and Biophysical Research Communications* **401**, 521–526.

109. Stephens, L., Milne, L., and Hawkins, P. (2008). Moving towards a better understanding of chemotaxis. *Current Biology* **18**, R485–R494.

110. Xie, L., Altindal, T., Chattopadhyay, S., and Wu, X.-L. (2011). Bacterial flagellum as a propeller and as a rudder for efficient chemotaxis. *Proceedings of the National Academy of Sciences of the United States of America* **108**, 2246–2251.

111. Terasawa, S., Fukuoka, H., Inoue, Y., Sagawa, T., Takahashi, H., and Ishijima, A. (2011). Coordinated reversal of flagellar motors on a single *Escherichia coli* cell. *Biophysical Journal* **100**, 2193–2200.

112. Boehm, A., Kaiser, M., Li, H., Spangler, C., Kasper, C. A., Ackermann, M., Kaever, V., Sourjik, V., Roth, V., and Jenal, U. (2010). Second messenger-mediated adjustment of bacterial swimming velocity. *Cell* **141**, 107–116.

113. Li, H., and Sourjik, V. (2011). Assembly and stability of flagellar motor in *Escherichia coli*. *Molecular Microbiology* **80**, 886-899.

114. Li, H., and Sourjik, V. (2011). Assembly and stability of flagellar motor in *Escherichia coli*. *Molecular Microbiology* **80**, 886–899.

115. Minamino, T., Imada, K., and Namba, K. (2008). Molecular motors of the bacterial flagella. *Current Opinion in Structural Biology* **18**, 693–701.

116. Macnab, R. M. (1999). The bacterial flagellum: Reversible rotary propellor and type III export apparatus. *Journal of Bacteriology* **181**, 7149–7153.

117. DePamphilis, M. L., and Adler, J. (1971). Fine structure and isolation of the hook-basal body complex of flagella from *Escherichia coli* and *Bacillus subtilis*. *Journal of Bacteriology* **105**, 384–395.

118. DePamphilis, M. L., and Adler, J. (1971). Attachment of flagellar basal bodies to the cell envelope: Specific attachment to the outer, lipopolysaccharide membrane, and the cyoplasmic membrane. *Journal of Bacteriology* **105**, 396–407.

119. DePamphilis, M. L., and Adler, J. (1971). Purification of intact flagella from *Escherichia coli* and *Bacillus subtilis*. *Journal of Bacteriology* **105**, 376–383.

120. Thomas, D. R., Francis, N. R., Xu, C., and DeRosier, D. J. (2006). The three-dimensional structure of the flagellar rotor from a clockwise-locked mutant of Salmonella enterica serovar Typhimurium. *Journal of Bacteriology* **188**, 7039–7048.

121. Suzuki, H., Yonekura, K., and Namba, K. (2004). Structure of the rotor of the bacterial flagellar motor revealed by electron cryomicroscopy and single-particle image analysis. *Journal of Molecular Biology* **337**, 105–113.

122. Murphy, G. E., Leadbetter, J. R., and Jensen, G. J. (2006). *In situ* structure of the complete Treponema primitia flagellar motor. *Nature* **442**, 1062–1064.

123. Chen, S., Beeby, M., Murphy, G. E., Leadbetter, J. R., Hendrixson, D. R., Briegel, A., Li, Z., Shi, J., Tocheva, E. I., Muller, A., Dobro, M. J., and Jensen, G. J. (2011). Structural diversity of bacterial flagellar motors. *The EMBO Journal* **30**, 2972–2981.

124. Berg, H. C. (2000). Constraints on models for the flagellar rotary motor. *Philosophical Transactions of the Royal Society of London. Series B: Biological Sciences* **355**, 491–501.

125. Magariyama, Y., Sugiyama, S., Muramoto, K., Maekawa, Y., Kawagishi, I., Imae, Y., and Kudo, S. (1994). Very fast flagellar rotation. *Nature* **371**, 752.

126. Muramoto, K., Kawagishi, I., Kudo, S., Magariyama, Y., Imae, Y., and Homma, M. (1995). High-speed rotation and speed stability of the sodium-driven flagellar motor in Vibrio alginolyticus. *Journal of Molecular Biology* **251**, 50–58.

127. Zhou, J., Lloyd, S. A., and Blair, D. F. (1998). Electrostatic interactions between rotor and stator in the bacterial flagellar motor. *Proceedings of the National Academy of Sciences of the United States of America* **95**, 6436–6441.

128. Francis, N. R., Irikura, V. M., Yamaguchi, S., DeRosier, D. J., and Macnab, R. M. (1992). Localization of the Salmonella typhimurium flagellar switch protein FliG to the cytoplasmic M-ring face of the basal body. *Proceedings of the National Academy of Sciences of the United States of America* **89**, 6304–6308.

129. Khan, S., Dapice, M., and Reese, T. S. (1988). Effects of mot gene expression on the structure of the flagellar motor. *Journal of Molecular Biology* **202**, 575–584.

130. Paul, K., Gonzalez-Bonet, G., Bilwes, A. M., Crane, B. R., and Blair, D. (2011). Architecture of the flagellar rotor. *The EMBO Journal* **30**, 2962–2971.

131. Erhardt, M., Singer, H. M., Wee, D. H., Keener, J. P., and Hughes, K. T. (2011). An infrequent molecular ruler controls flagellar hook length in Salmonella enterica. *The EMBO Journal* **30**, 2948–2961.

132. Muskotál, A., Seregélyes, C., Sebestyén, A., and Vonderviszt, F. (2010). Structural basis for stabilization of the hypervariable D3 domain of Salmonella flagellin upon filament formation. *Journal of Molecular Biology* **403**, 607–615.

133. Larsen, S. H., Adler, J., Gargus, J. J., and Hogg, R. W. (1974). Chemomechanical coupling without ATP: The source of energy for motility and chemotaxis in bacteria. *Proceedings of the National Academy of Sciences of the United States of America* **71**, 1239–1243.

134. Kojima, S., and Blair, D. F. (2001). Conformational change in the stator of the bacterial flagellar motor. *Biochemistry* **40**, 13041–13050.

135. Morimoto, Y. V., Nakamura, S., Kami-ike, N., Namba, K., and Minamino, T. (2010). Charged residues in the cytoplasmic loop of MotA are required for stator assembly into the bacterial flagellar motor. *Molecular Microbiology* **78**, 1117–1129.

136. Zhou, J., Sharp, L. L., Tang, H. L., Lloyd, S. A., Billings, S., Braun, T. F., and Blair, D. F. (1998). Function of protonatable residues in the flagellar motor of *Escherichia coli*: A critical role for Asp 32 of MotB. *Journal of Bacteriology* **180**, 2729–2735.

137. Fukuoka, H., Wada, T., Kojima, S., Ishijima, A., and Homma, M. (2009). Sodium-dependent dynamic assembly of membrane complexes in sodium-driven flagellar motors. *Molecular Microbiology* **71**, 825–835.

138. Terashima, H., Kojima, S., and Homma, M. (2010). Functional transfer of an essential aspartate for the ion-binding site in the stator proteins of the bacterial flagellar motor. *Journal of Molecular Biology* **397**, 689–696.

139. Ibuki, T., Imada, K., Minamino, T., Kato, T., Miyata, T., and Namba, K. (2011). Common architecture of the flagellar type III protein export apparatus and F- and V-type ATPases. *Nature Structural and Molecular Biology* **18**, 277–282.

140. Nakamura, S., Kami-ike, N., Yokota, J.-i. P., Minamino, T., and Namba, K. (2010). Evidence for symmetry in the elementary process of bidirectional torque generation by the bacterial flagellar motor. *Proceedings of the National Academy of Sciences of the United States of America* **107**, 17616–17620.

141. Lee, L. K., Ginsburg, M. A., Crovace, C., Donohoe, M., and Stock, D. (2010). Structure of the torque ring of the flagellar motor and the molecular basis for rotational switching. *Nature* **466**, 996–1000.

142. Watari, N., and Larson, R. G. (2010). The Hydrodynamics of a Run-and-Tumble Bacterium Propelled by Polymorphic Helical Flagella. *Biophysical Journal* **98**, 12–17.

143. Imada, K., Minamino, T., Tahara, A., and Namba, K. (2007). Structural similarity between the flagellar type III ATPase FliI and F_1-ATPase subunits. *Proceedings of the National Academy of Sciences of the United States of America* **104**, 485–90.

144. Minamino, T., and Namba, K. (2008). Distinct roles of the FliI ATPase and proton motive force in bacterial flagellar protein export. *Nature* **451**, 485–8.

145. Paul, K., Erhardt, M., Hirano, T., Blair, D. F., and Hughes, K. T. (2008). Energy source of flagellar type III secretion. *Nature* **451**, 489–92.

146. Paulick, A., Koerdt, A., Lassak, J., Huntley, S., Wilms, I., Narberhaus, F., and Thormann, K. M. (2009). Two different stator systems drive a single polar flagellum in Shewanella oneidensis MR-1. *Molecular Microbiology* **71**, 836–50.

147. Leake, M. C., Chandler, J. H., Wadhams, G. H., Bai, F., Berry, R. M., and Armitage, J. P. (2006). Stoichiometry and turnover in single, functioning membrane protein complexes. *Nature* **443**, 355–8.

148. Fukuoka, H., Inoue, Y., Terasawa, S., Takahashi, H., and Ishijima, A. (2010). Exchange of rotor components in functioning bacterial flagellar motor. *Biochemical and Biophysical Research Communications* **394**, 130–135.

149. Lowe, M. A., Holt, S. C., and Eisenstein, B. I. (1987). Immunoelectron microscopic analysis of elongation of type 1 fimbriae in Escherichia coli. *Journal of Bacteriology* **169**, 157–63.

150. Berry, R. M., and Berg, H. C. (1996). Torque generated by the bacterial flagellar motor close to stall. *Biophysical Journal* **71**, 3501–3510.

151. Berry, R. M., and Berg, H. C. (1997). Absence of a barrier to backwards rotation of the bacterial flagellar motor demonstrated with optical tweezers. *Proceedings of the National Academy of Sciences of the United States of America* **94**, 14433–14437.

152. Berry, R. M., and Berg, H. C. (1999). Torque generated by the flagellar motor of *Escherichia coli* while driven backward. *Biophysical Journal* **76**, 580–587.

153. Bai, F., Lo, C. J., Berry, R. M., and Xing, J. (2009). Model studies of the dynamics of bacterial flagellar motors. *Biophysical Journal* **96**, 3154–3167.

154. Meacci, G., Lan, G., and Tu, Y. (2011). Dynamics of the bacterial flagellar motor: The effects of stator compliance, back steps, temperature, and rotational asymmetry. *Biophysical Journal* **100**, 1986–1995.

155. Axelsen, K. B., and Palmgren, M. G. (1998). Evolution of substrate specificities in the P-type ATPase superfamily. *Journal of Molecular Evolution* **46**, 84–101.

156. Bublitz, M., Poulsen, H., Morth, J. P., and Nissen, P. (2010). In and out of the cation pumps: P-type ATPase structure revisited. *Current Opinion in Structural Biology* **20**, 431–439.

157. Allen, K. N., and Dunaway-Mariano, D. (2004). Phosphoryl group transfer: evolution of a catalytic scaffold. *Trends in Biochemical Sciences* **29**, 495–503.

158. Rao, V. B., and Black, L. W. (2010). Structure and assembly of bacteriophage T4 head. *Virology Journal* **7**, 356.

159. Black, L. W. (1989). DNA packaging in dsDNA bacteriophages. *Annual Review of Microbiology* **43**, 267–292.

160. Earnshaw, W. C., and Casjens, S. R. (1980). DNA packaging by the double-stranded DNA bacteriophages. *Cell* **21**, 319–331.

161. Smith, D. E., Tans, S. J., Smith, S. B., Grimes, S., Anderson, D. L., and Bustamante, C. (2001). The bacteriophage phis29 portal motor can package DNA against a large internal force. *Nature* **413**, 748–752.

162. Hays, J. B., Magar, M. E., and Zimm, B. H. (1969). Persistence length of DNA. *Biopolymers* **8**, 531–536.

163. Fuller, D. N., Raymer, D. M., Rickgauer, J. P., Robertson, R. M., Catalano, C. E., Anderson, D. L., Grimes, S., and Smith, D. E. (2007). Measurements of single DNA molecule packaging dynamics in bacteriophage lambda reveal high forces, high motor processivity, and capsid transformations. *Journal of Molecular Biology* **373**, 1113–1122.

164. Fuller, D. N., Raymer, D. M., Kottadiel, V. I., Rao, V. B., and Smith, D. E. (2007). Single phage T4 DNA packaging motors exhibit large force generation, high velocity, and dynamic variability. *Proceedings of the National Academy of Sciences of the United States of America* **104**, 16868–16873.

165. Rao, V. B., and Feiss, M. (2008). The bacteriophage DNA packaging motor. *Annual Review of Genetics* **42**, 647–681.

166. Casjens, S. R., Gilcrease, E. B., Winn-Stapley, D. A., Schicklmaier, P., Schmieger, H., Pedulla, M. L., Ford, M. E., Houtz, J. M., Hatfull, G. F., and Hendrix, R. W. (2005). The generalized transducing Salmonella bacteriophage ES18: complete genome sequence and DNA packaging strategy. *Journal of Bacteriology* **187**, 1091–1104.

167. Draper, B., and Rao, V. B. (2007). An ATP hydrolysis sensor in the DNA packaging motor from bacteriophage T4 suggests an inchworm-type translocation mechanism. *Journal of Molecular Biology* **369**, 79–94.

168. Mitchell, M. S., Matsuzaki, S., Imai, S., and Rao, V. B. (2002). Sequence analysis of bacteriophage T4 DNA packaging/terminase genes 16 and 17 reveals a common ATPase center in the large subunit of viral terminases. *Nucleic Acids Research* **30**, 4009–4021.

169. Sun, J., Cai, Y., Moll, W.-D., and Guo, P. (2006). Controlling bacteriophage phi29 DNA-packaging motor by addition or discharge of a peptide at N-terminus of connector protein that interacts with pRNA. *Nucleic Acids Research* **34**, 5482–5490.

170. Kitamura, A., Jardine, P. J., Anderson, D. L., Grimes, S., and Matsuo, H. (2008). Analysis of intermolecular base pair formation of prohead RNA

of the phage phi29 DNA packaging motor using NMR spectroscopy. *Nucleic Acids Research* **36**, 839–848.

171. Xiao, F., Moll, W. D., Guo, S., and Guo, P. (2005). Binding of pRNA to the N-terminal 14 amino acids of connector protein of bacteriophage phi29. *Nucleic Acids Research* **33**, 2640–2649.

172. Atz, R., Ma, S., Gao, J., Anderson, D. L., and Grimes, S. (2007). Alanine scanning and Fe-BABE probing of the bacteriophage o29 prohead RNA-connector interaction. *Journal of Molecular Biology* **369**, 239–248.

173. Guo, P., Peterson, C., and Anderson, D. (1987). Prohead and DNA-gp3-dependent ATPase activity of the DNA packaging protein gp16 of bacteriophage phi29. *Journal of Molecular Biology* **197**, 229–236.

174. Morita, M., Tasaka, M., and Fujisawa, H. (1993). DNA packaging ATPase of bacteriophage T3. *Virology* **193**, 748–752.

175. Chemla, Y. R., Aathavan, K., Michaelis, J., Grimes, S., Jardine, P. J., Anderson, D. L., and Bustamante, C. (2005). Mechanism of force generation of a viral DNA packaging motor. *Cell* **122**, 683–692.

176. Simpson, A. A., Tao, Y., Leiman, P. G., Badasso, M. O., He, Y., Jardine, P. J., Olson, N. H., Morais, M. C., Grimes, S., Anderson, D. L., Baker, T. S., and Rossmann, M. G. (2000). Structure of the bacteriophage phi29 DNA packaging motor. *Nature* **408**, 745–750.

177. Lebedev, A. A., Krause, M. H., Isidro, A. L., Vagin, A. A., Orlova, E. V., Turner, J., Dodson, E. J., Tavares, P., and Antson, A. A. (2007). Structural framework for DNA translocation via the viral portal protein. *The EMBO Journal* **26**, 1984–1994.

178. Orlova, E. V., Gowen, B., Droge, A., Stiege, A., Weise, F., Lurz, R., van Heel, M., and Tavares, P. (2003). Structure of a viral DNA gatekeeper at 10 A resolution by cryo-electron microscopy. *The EMBO Journal* **22**, 1255–1262.

179. Leiman, P. G., Chipman, P. R., Kostyuchenko, V. A., Mesyanzhinov, V. V., and Rossmann, M. G. (2004). Three-dimensional rearrangement of proteins in the tail of bacteriophage T4 on infection of its host. *Cell* **118**, 419–429.

180. Rao, V. B., and Mitchell, M. S. (2001). The N-terminal ATPase site in the large terminase protein gp17 is critically required for DNA packaging in bacteriophage T4. *Journal of Molecular Biology* **314**, 401–411.

181. Grimes, S., Ma, S., Gao, J., Atz, R., and Jardine, P. J. (2011). Role of [phi]29 connector channel loops in late-stage DNA packaging. *Journal of Molecular Biology* **410**, 50–59.

182. Becker, A., and Murialdo, H. (1990). Bacteriophage lambda DNA: the beginning of the end. *Journal of Bacteriology* **172**, 2819–2824.

183. de Beer, T., Fang, J., Ortega, M., Yang, Q., Maes, L., Duffy, C., Berton, N., Sippy, J., Overduin, M., Feiss, M., and Catalano, C. E. (2002). Insights into specific DNA recognition during the assembly of a viral genome packaging machine. *Molecular Cell* **9**, 981–991.

184. Chai, S., Lurz, R., and Alonso, J. C. (1995). The small subunit of the terminase enzyme of *Bacillus subtilis* bacteriophage SPP1 forms a specialized nucleoprotein complex with the packaging initiation region. *Journal of Molecular Biology* **252**, 386–398.

185. Lin, H., Simon, M. N., and Black, L. W. (1997). Purification and characterization of the small subunit of phage T4 terminase, gp16, required for DNA packaging. *The Journal of Biological Chemistry* **272**, 3495–3501.

186. Nemecek, D., Gilcrease, E. B., Kang, S., Prevelige Jr, P. E., Casjens, S., and Thomas Jr, G. J. (2007). Subunit conformations and assembly states of a DNA-translocating motor: The terminase of bacteriophage P22. *Journal of Molecular Biology* **374**, 817–836.

187. Frackman, S., Siegele, D. A., and Feiss, M. (1985). The terminase of bacteriophage lambda. Functional domains for cosB binding and multimer assembly. *Journal of Molecular Biology* **183**, 225–238.

188. Gao, S., and Rao, V. B. (2011). Specificity of interactions among the DNA-packaging machine components of T4-related bacteriophages. *The Journal of Biological Chemistry* **286**, 3944–3956.

189. Mitchell, M. S., and Rao, V. B. (2006). Functional analysis of the bacteriophage T4 DNA-packaging ATPase motor. *The Journal of Biological Chemistry* **281**, 518–527.

190. Story, R. M., and Steitz, T. A. (1992). Structure of the recA protein-ADP complex. *Nature* **355**, 374–376.

191. Story, R. M., Weber, I. T., and Steitz, T. A. (1992). The structure of the *E. coli* recA protein monomer and polymer. *Nature* **355**, 318–325.

192. Gorbalenya, A. E., and Koonin, E. V. (1993). Helicases: amino acid sequence comparisons and structure-function relationships. *Current Opinion in Structural Biology* **3**, 419–429.

193. Duffy, C., and Feiss, M. (2002). The large subunit of bacteriophage lambda's terminase plays a role in DNA translocation and packaging termination. *Journal of Molecular Biology* **316**, 547–561.

194. Sun, S., Kondabagil, K., Draper, B., Alam, T. I., Bowman, V. D., Zhang, Z., Hegde, S., Fokine, A., Rossmann, M. G., and Rao, V. B. (2008). The

structure of the phage T4 DNA packaging motor suggests a mechanism dependent on electrostatic forces. *Cell* **135**, 1251–1262.

195. Yu, J., Moffitt, J., Hetherington, C. L., Bustamante, C., and Oster, G. (2010). Mechanochemistry of a viral DNA packaging motor. *Journal of Molecular Biology* **400**, 186–203.

196. Chemla, Y. R., Aathavan, K., Michaelis, J., Grimes, S., Jardine, P. J., Anderson, D. L., and Bustamante, C. (2005). Mechanism of force generation of a viral DNA packaging motor. *Cell* **122**, 683–692.

197. Fuller, D. N., Rickgauer, J. P., Jardine, P. J., Grimes, S., Anderson, D. L., and Smith, D. E. (2007). Ionic effects on viral DNA packaging and portal motor function in bacteriophage phi29. *Proceedings of the National Academy of Sciences of the United States of America* **104**, 11245–11250.

198. Rickgauer, J. P., Fuller, D. N., Grimes, S., Jardine, P. J., Anderson, D. L., and Smith, D. E. (2007). Portal motor velocity and capsid pressure during viral DNA packaging in bacteriophage phi29. *Biophysical Journal*; [Epub ahead of print]. PMID: 17827233

199. Rickgauer, J. P., Fuller, D. N., Grimes, S., Jardine, P. J., Anderson, D. L., and Smith, D. E. (2008). Portal motor velocity and internal force resisting viral DNA packaging in bacteriophage phi29. *Biophysical Journal* **94**, 159–167.

200. Smith, D. E., and Tsay, J. M. (2009). Single-molecule studies of viral DNA packaging with optical tweezers: Molecular motor function and DNA confinement. **96**, 15a–16a.

201. Purohit, P. K., Inamdar, M. M., Grayson, P. D., Squires, T. M., Kondev, J., and Phillips, R. (2005). Forces during bacteriophage DNA packaging and ejection. *Biophysical Journal* **88**, 851–866.

202. Su, T., Cui, Y., Zhang, X., Liu, X., Yue, J., Liu, N., and Jiang, P. (2006). Constructing a novel Nanodevice powered by [delta]-free F_0F_1 -ATPase. *Biochemical and Biophysical Research Communications* **350**, 1013–1018.

203. Vale, R. D., Reese, T. S., and Sheetz, M. P. (1985). Identification of a novel force-generating protein, kinesin, involved in microtubule-based motility. *Cell* **42**, 39–50.

204. Vale, R. D., and Milligan, R. A. (2000). The way things move: looking under the hood of molecular motor proteins. *Science* **288**, 88–95.

205. Vale, R. D., Funatsu, T., Pierce, D. W., Romberg, L., Harada, Y., and Yanagida, T. (1996). Direct observation of single kinesin molecules moving along microtubules. *Nature* **380**, 451–453.

206. Kojima, H., Muto, E., Higuchi, H., and Yanagida, T. (1997). Mechanics of single kinesin molecules measured by optical trapping nanometry. *Biophysical Journal* **73**, 2012–2022.

207. Hirokawa, N. (1998). Kinesin and dynein superfamily proteins and the mechanism of organelle transport. *Science* **279**, 519–526.

208. Hirokawa, N., Pfister, K. K., Yorifuji, H., Wagner, M. C., Brady, S. T., and Bloom, G. S. (1989). Submolecular domains of bovine brain kinesin identified by electron microscopy and monoclonal antibody decoration. *Cell* **56**, 867–878.

209. Hirokawa, N., and Takemura, R. (2005). Molecular motors and mechanisms of directional transport in neurons. *Nature Reviews Neuroscience* **6**, 201–214.

210. Lawrence, C. J., Dawe, R. K., Christie, K. R., Cleveland, D. W., Dawson, S. C., Endow, S. A., Goldstein, L. S., Goodson, H. V., Hirokawa, N., Howard, J., Malmberg, R. L., McIntosh, J. R., Miki, H., Mitchison, T. J., Okada, Y., Reddy, A. S., Saxton, W. M., Schliwa, M., Scholey, J. M., Vale, R. D., Walczak, C. E., and Wordeman, L. (2004). A standardized kinesin nomenclature. *The Journal of Cell Biology* **167**, 19–22.

211. Hirokawa, N., and Noda, Y. (2008). Intracellular transport and kinesin superfamily proteins, KIFs: structure, function, and dynamics. *Physiological Reviews* **88**, 1089–1118.

212. Kikkawa, M., Sablin, E. P., Okada, Y., Yajima, H., Fletterick, R. J., and Hirokawa, N. (2001). Switch-based mechanism of kinesin motors. *Nature* **411**, 439–445.

213. Nitta, R., Kikkawa, M., Okada, Y., and Hirokawa, N. (2004). KIF1A alternately uses two loops to bind microtubules. *Science* **305**, 678–683.

214. Diefenbach, R. J., Mackay, J. P., Armati, P. J., and Cunningham, A. L. (1998). The C-terminal region of the stalk domain of ubiquitous human kinesin heavy chain contains the binding site for kinesin light chain. *Biochemistry* **37**, 16663–16670.

215. Kanai, Y., Dohmae, N., and Hirokawa, N. (2004). Kinesin transports RNA: Isolation and characterization of an RNA-transporting granule. *Neuron* **43**, 513–525.

216. Case, R. B., Pierce, D. W., Hom-Booher, N., Hart, C. L., and Vale, R. D. (1997). The directional preference of kinesin motors is specified by an element outside of the motor catalytic domain. *Cell* **90**, 959–966.

217. Case, R. B., Rice, S., Hart, C. L., Ly, B., and Vale, R. D. (2000). Role of the kinesin neck linker and catalytic core in microtubule-based motility. *Current Biology* **10**, 157–160.

218. Henningsen, U., and Schliwa, M. (1997). Reversal in the direction of movement of a molecular motor. *Nature* **389**, 93–96.

219. Rice, S., Lin, A. W., Safer, D., Hart, C. L., Naber, N., Carragher, B. O., Cain, S. M., Pechatnikova, E., Wilson-Kubalek, E. M., Whittaker, M., Pate, E., Cooke, R., Taylor, E. W., Milligan, R. A., and Vale, R. D. (1999). A structural change in the kinesin motor protein that drives motility. *Nature* **402**, 778–784.

220. Rosenfeld, S. S., Jefferson, G. M., and King, P. H. (2001). ATP reorients the neck linker of kinesin in two sequential steps. *The Journal of Biological Chemistry* **276**, 40167–40174.

221. Miki, H., Okada, Y., and Hirokawa, N. (2005). Analysis of the kinesin superfamily: insights into structure and function. *Trends in Cell Biology* **15**, 467–476.

222. Brady, S. T., Pfister, K. K., and Bloom, G. S. (1990). A monoclonal antibody against kinesin inhibits both anterograde and retrograde fast axonal transport in squid axoplasm. *Proceedings of the National Academy of Sciences of the United States of America* **87**, 1061–1065.

223. Holzinger, A., and Lutz-Meindl, U. (2002). Kinesin-like proteins are involved in postmitotic nuclear migration of the unicellular green alga Micrasterias denticulata. *Cell Biology International* **26**, 689–697.

224. Palacios, I. M., and St Johnston, D. (2002). Kinesin light chain-independent function of the kinesin heavy chain in cytoplasmic streaming and posterior localisation in the *Drosophila* oocyte. *Development* **129**, 5473–5485.

225. Pfister, K. K., Wagner, M. C., Stenoien, D. L., Brady, S. T., and Bloom, G. S. (1989). Monoclonal antibodies to kinesin heavy and light chains stain vesicle-like structures, but not microtubules, in cultured cells. *The Journal of Cell Biology* **108**, 1453–1463.

226. Gauger, A. K., and Goldstein, L. S. (1993). The *Drosophila* kinesin light chain. Primary structure and interaction with kinesin heavy chain. *The Journal of Biological Chemistry* **268**, 13657–13666.

227. Muresan, V., Abramson, T., Lyass, A., Winter, D., Porro, E., Hong, F., Chamberlin, N. L., and Schnapp, B. J. (1998). KIF3C and KIF3A form a novel neuronal heteromeric kinesin that associates with membrane vesicles. *Molecular Biology of the Cell* **9**, 637–652.

228. Setou, M., Nakagawa, T., Seog, D. H., and Hirokawa, N. (2000). Kinesin superfamily motor protein KIF17 and mLin-10 in NMDA receptor-containing vesicle transport. *Science* **288**, 1796–1802.

229. Yamazaki, H., Nakata, T., Okada, Y., and Hirokawa, N. (1995). KIF3A/B: a heterodimeric kinesin superfamily protein that works as a microtubule

plus end-directed motor for membrane organelle transport. *The Journal of Cell Biology* **130**, 1387–1399.

230. Henson, J. H., Cole, D. G., Roesener, C. D., Capuano, S., Mendola, R. J., and Scholey, J. M. (1997). The heterotrimeric motor protein kinesin-II localizes to the midpiece and flagellum of sea urchin and sand dollar sperm. *Cell Motility and the Cytoskeleton* **38**, 29–37.

231. Wedaman, K. P., Meyer, D. W., Rashid, D. J., Cole, D. G., and Scholey, J. M. (1996). Sequence and submolecular localization of the 115-kD accessory subunit of the heterotrimeric kinesin-II (KRP85/95) complex. *The Journal of Cell Biology* **132**, 371–380.

232. Yamazaki, H., Nakata, T., Okada, Y., and Hirokawa, N. (1996). Cloning and characterization of KAP3: a novel kinesin superfamily-associated protein of KIF3A/3B. *Proceedings of the National Academy of Sciences of the United States of America* **93**, 8443–8448.

233. Durocher, D., Henckel, J., Fersht, A. R., and Jackson, S. P. (1999). The FHA domain is a modular phosphopeptide recognition motif. *Molecular Cell* **4**, 387–394.

234. Westerholm-Parvinen, A., Vernos, I., and Serrano, L. (2000). Kinesin subfamily UNC104 contains a FHA domain: boundaries and physicochemical characterization. *FEBS Letters* **486**, 285–290.

235. Lee, J. R., Shin, H., Choi, J., Ko, J., Kim, S., Lee, H. W., Kim, K., Rho, S. H., Lee, J. H., Song, H. E., Eom, S. H., and Kim, E. (2004). An intramolecular interaction between the FHA domain and a coiled coil negatively regulates the kinesin motor KIF1A. *The EMBO Journal* **23**, 1506–1515.

236. Hall, D. H., and Hedgecock, E. M. (1991). Kinesin-related gene unc-104 is required for axonal transport of synaptic vesicles in *C. elegans*. *Cell* **65**, 837–847.

237. Okada, Y., Yamazaki, H., Sekine-Aizawa, Y., and Hirokawa, N. (1995). The neuron-specific kinesin superfamily protein KIF1A is a unique monomeric motor for anterograde axonal transport of synaptic vesicle precursors. *Cell* **81**, 769–780.

238. Yonekawa, Y., Harada, A., Okada, Y., Funakoshi, T., Kanai, Y., Takei, Y., Terada, S., Noda, T., and Hirokawa, N. (1998). Defect in synaptic vesicle precursor transport and neuronal cell death in KIF1A motor protein-deficient mice. *The Journal of Cell Biology* **141**, 431–441.

239. Nangaku, M., Sato-Yoshitake, R., Okada, Y., Noda, Y., Takemura, R., Yamazaki, H., and Hirokawa, N. (1994). KIF1B, a novel microtubule plus end-directed monomeric motor protein for transport of mitochondria. *Cell* **79**, 1209–1220.

240. Dorner, C., Ullrich, A., Haring, H. U., and Lammers, R. (1999). The kinesin-like motor protein KIF1C occurs in intact cells as a dimer and associates with proteins of the 14-3-3 family. *The Journal of Biological Chemistry* **274**, 33654–33660.

241. Miki, H., Setou, M., Kaneshiro, K., and Hirokawa, N. (2001). All kinesin superfamily protein, KIF, genes in mouse and human. *Proceedings of the National Academy of Sciences of the United States of America* **98**, 7004–7011.

242. Vale, R. D., and Fletterick, R. J. (1997). The design plan of kinesin motors. *Annual Review of Cell and Developmental Biology* **13**, 745–777.

243. Dagenbach, E. M., and Endow, S. A. (2004). A new kinesin tree. *Journal of Cell Science* **117**, 3–7.

244. Lawrence, C. J., Malmberg, R. L., Muszynski, M. G., and Dawe, R. K. (2002). Maximum likelihood methods reveal conservation of function among closely related kinesin families. *Journal of Molecular Evolution* **54**, 42–53.

245. Sekine, Y., Okada, Y., Noda, Y., Kondo, S., Aizawa, H., Takemura, R., and Hirokawa, N. (1994). A novel microtubule-based motor protein (KIF4) for organelle transports, whose expression is regulated developmentally. *The Journal of Cell Biology* **127**, 187–201.

246. Vernos, I., Raats, J., Hirano, T., Heasman, J., Karsenti, E., and Wylie, C. (1995). Xklp1, a chromosomal Xenopus kinesin-like protein essential for spindle organization and chromosome positioning. *Cell* **81**, 117–127.

247. Wang, S. Z., and Adler, R. (1995). Chromokinesin: a DNA-binding, kinesin-like nuclear protein. *The Journal of Cell Biology* **128**, 761–768.

248. Williams, B. C., Riedy, M. F., Williams, E. V., Gatti, M., and Goldberg, M. L. (1995). The *Drosophila* kinesin-like protein KLP3A is a midbody component required for central spindle assembly and initiation of cytokinesis. *The Journal of Cell Biology* **129**, 709–723.

249. Le Guellec, R., Paris, J., Couturier, A., Roghi, C., and Philippe, M. (1991). Cloning by differential screening of a Xenopus cDNA that encodes a kinesin-related protein. *Molecular and Cellular Biology* **11**, 3395–3398.

250. Blangy, A., Lane, H. A., d'Herin, P., Harper, M., Kress, M., and Nigg, E. A. (1995). Phosphorylation by p34cdc2 regulates spindle association of human Eg5, a kinesin-related motor essential for bipolar spindle formation *in vivo*. *Cell* **83**, 1159–1169.

251. Larson, A. G., Naber, N., Cooke, R., Pate, E., and Rice, S. E. (2010). The conserved L5 Loop establishes the pre-powerstroke conformation of the kinesin-5 motor, Eg5. *Biophysical Journal* **98**, 2619–2627.

252. Hemsley, R., McCutcheon, S., Doonan, J., and Lloyd, C. (2001). P34(cdc2) kinase is associated with cortical microtubules from higher plant protoplasts. *FEBS Letters* **508**, 157–161.

253. Hemerly, A. S., Ferreira, P., de Almeida Engler, J., Van Montagu, M., Engler, G., and Inze, D. (1993). cdc2a expression in Arabidopsis is linked with competence for cell division. *Plant Cell* **5**, 1711–1723.

254. Cole, D. G., Saxton, W. M., Sheehan, K. B., and Scholey, J. M. (1994). A "slow" homotetrameric kinesin-related motor protein purified from *Drosophila* embryos. *The Journal of Biological Chemistry* **269**, 22913–22916.

255. Enos, A. P., and Morris, N. R. (1990). Mutation of a gene that encodes a kinesin-like protein blocks nuclear division in *A. nidulans*. *Cell* **60**, 1019–1027.

256. Kikkawa, M., Okada, Y., and Hirokawa, N. (2000). 15 A resolution model of the monomeric kinesin motor, KIF1A. *Cell* **100**, 241–252.

257. Adams, R. R., Tavares, A. A., Salzberg, A., Bellen, H. J., and Glover, D. M. (1998). Pavarotti encodes a kinesin-like protein required to organize the central spindle and contractile ring for cytokinesis. *Genes and Development* **12**, 1483–1494.

258. Hill, E., Clarke, M., and Barr, F. A. (2000). The Rab6-binding kinesin, Rab6-KIFL, is required for cytokinesis. *The EMBO Journal* **19**, 5711–5719.

259. Nislow, C., Lombillo, V. A., Kuriyama, R., and McIntosh, J. R. (1992). A plus-end-directed motor enzyme that moves antiparallel microtubules *in vitro* localizes to the interzone of mitotic spindles. *Nature* **359**, 543–547.

260. Weaver, B. A., Bonday, Z. Q., Putkey, F. R., Kops, G. J., Silk, A. D., and Cleveland, D. W. (2003). Centromere-associated protein-E is essential for the mammalian mitotic checkpoint to prevent aneuploidy due to single chromosome loss. *The Journal of Cell Biology* **162**, 551–563.

261. Miller, R. K., Heller, K. K., Frisen, L., Wallack, D. L., Loayza, D., Gammie, A. E., and Rose, M. D. (1998). The kinesin-related proteins, Kip2p and Kip3p, function differently in nuclear migration in yeast. *Molecular Biology of the Cell* **9**, 2051–2068.

262. Peters, C., Brejc, K. a., Belmont, L., Bodey, A. J., Lee, Y., Yu, M., Guo, J., Sakowicz, R., Hartman, J., and Moores, C. A. (2010). Insight into the molecular mechanism of the multitasking kinesin-8 motor. *The EMBO Journal* **29**, 3437–3447.

263. DeZwaan, T. M., Ellingson, E., Pellman, D., and Roof, D. M. (1997). Kinesin-related KIP3 of Saccharomyces cerevisiae is required for

a distinct step in nuclear migration. *The Journal of Cell Biology* **138**, 1023–1040.

264. Cottingham, F. R., and Hoyt, M. A. (1997). Mitotic spindle positioning in Saccharomyces cerevisiae is accomplished by antagonistically acting microtubule motor proteins. *The Journal of Cell Biology* **138**, 1041–1053.

265. Pereira, A. J., Dalby, B., Stewart, R. J., Doxsey, S. J., and Goldstein, L. S. (1997). Mitochondrial association of a plus end-directed microtubule motor expressed during mitosis in Drosophila. *The Journal of Cell Biology* **136**, 1081–1090.

266. West, R. R., Malmstrom, T., and McIntosh, J. R. (2002). Kinesins klp5(+) and klp6(+) are required for normal chromosome movement in mitosis. *Journal of Cell Science* **115**, 931–940.

267. Stumpff, J., von Dassow, G., Wagenbach, M., Asbury, C., and Wordeman, L. (2008). The kinesin-8 motor KIF18A suppresses kinetochore movements to control mitotic chromosome alignment. *Developmental Cell* **14**, 252–262.

268. Bernstein, M., Beech, P. L., Katz, S. G., and Rosenbaum, J. L. (1994). A new kinesin-like protein (Klp1) localized to a single microtubule of the Chlamydomonas flagellum. *The Journal of Cell Biology* **125**, 1313–1326.

269. Demonchy, R., Blisnick, T., Deprez, C., Toutirais, G., Loussert, C., Marande, W., Grellier, P., Bastin, P., and Kohl, L. (2009). Kinesin-9 family members perform separate functions in the trypanosome flagellum. *The Journal of Cell Biology* **187**, 615–622.

270. Lechtreck, K. F., and Witman, G. B. (2007). Chlamydomonas reinhardtii hydin is a central pair protein required for flagellar motility. *The Journal of Cell Biology* **176**, 473–482.

271. Dawe, H. R., Shaw, M. K., Farr, H., and Gull, K. (2007). The hydrocephalus inducing gene product, Hydin, positions axonemal central pair microtubules. *BMC Biology* **5**, 33.

272. Yokoyama, R., O'Toole, E., Ghosh, S., and Mitchell, D. R. (2004). Regulation of flagellar dynein activity by a central pair kinesin. *Proceedings of the National Academy of Sciences of the United States of America* **101**, 17398–17403.

273. Levesque, A. A., and Compton, D. A. (2001). The chromokinesin Kid is necessary for chromosome arm orientation and oscillation, but not congression, on mitotic spindles. *The Journal of Cell Biology* **154**, 1135–1146.

274. Yajima, J., Edamatsu, M., Watai-Nishii, J., Tokai-Nishizumi, N., Yamamoto, T., and Toyoshima, Y. Y. (2003). The human chromokinesin Kid is a plus end-directed microtubule-based motor. *The EMBO Journal* **22**, 1067–1074.

275. Lillie, S. H., and Brown, S. S. (1994). Immunofluorescence localization of the unconventional myosin, Myo2p, and the putative kinesin-related protein, Smy1p, to the same regions of polarized growth in Saccharomyces cerevisiae. *The Journal of Cell Biology* **125**, 825–842.

276. Wolf, F. W., Hung, M. S., Wightman, B., Way, J., and Garriga, G. (1998). vab-8 is a key regulator of posteriorly directed migrations in C. elegans and encodes a novel protein with kinesin motor similarity. *Neuron* **20**, 655–666.

277. Zhou, R., Niwa, S., Homma, N., Takei, Y., and Hirokawa, N. (2009). KIF26A is an unconventional kinesin and regulates GDNF-Ret signaling in enteric neuronal development. *Cell* **139**, 802–813.

278. Beningo, K. A., Lillie, S. H., and Brown, S. S. (2000). The yeast kinesin-related protein Smy1p exerts its effects on the class V myosin Myo2p via a physical interaction. *Molecular Biology of the Cell* **11**, 691–702.

279. Lillie, S. H., and Brown, S. S. (1998). Smy1p, a kinesin-related protein that does not require microtubules. *The Journal of Cell Biology* **140**, 873–883.

280. Lee, Y. R., and Liu, B. (2000). Identification of a phragmoplast-associated kinesin-related protein in higher plants. *Current Biology* **10**, 797–800.

281. Lee, Y. R., Li, Y., and Liu, B. (2007). Two Arabidopsis phragmoplast-associated kinesins play a critical role in cytokinesis during male gametogenesis. *Plant Cell* **19**, 2595–2605.

282. Muller, S., Han, S., and Smith, L. G. (2006). Two kinesins are involved in the spatial control of cytokinesis in Arabidopsis thaliana. *Current Biology* **16**, 888–894.

283. Boleti, H., Karsenti, E., and Vernos, I. (1996). Xklp2, a novel Xenopus centrosomal kinesin-like protein required for centrosome separation during mitosis. *Cell* **84**, 49–59.

284. Buster, D. W., Baird, D. H., Yu, W., Solowska, J. M., Chauviere, M., Mazurek, A., Kress, M., and Baas, P. W. (2003). Expression of the mitotic kinesin KIF15 in postmitotic neurons: implications for neuronal migration and development. *Journal of Neurocytology* **32**, 79–96.

285. Noda, Y., Sato-Yoshitake, R., Kondo, S., Nangaku, M., and Hirokawa, N. (1995). KIF2 is a new microtubule-based anterograde motor that transports membranous organelles distinct from those carried

by kinesin heavy chain or KIF3A/B. *The Journal of Cell Biology* **129**, 157–167.

286. Desai, A., Verma, S., Mitchison, T. J., and Walczak, C. E. (1999). Kin I kinesins are microtubule-destabilizing enzymes. *Cell* **96**, 69–78.

287. Homma, N., Takei, Y., Tanaka, Y., Nakata, T., Terada, S., Kikkawa, M., Noda, Y., and Hirokawa, N. (2003). Kinesin superfamily protein 2A (KIF2A) functions in suppression of collateral branch extension. *Cell* **114**, 229–239.

288. Helenius, J., Brouhard, G., Kalaidzidis, Y., Diez, S., and Howard, J. (2006). The depolymerizing kinesin MCAK uses lattice diffusion to rapidly target microtubule ends. *Nature* **441**, 115–119.

289. Gorbsky, G. J. (2004). Mitosis: MCAK under the aura of Aurora B. *Current Biology* **14**, R346–R348.

290. Holmfeldt, P., Zhang, X., Stenmark, S., Walczak, C. E., and Gullberg, M. (2005). CaMKIIgamma-mediated inactivation of the Kin I kinesin MCAK is essential for bipolar spindle formation. *The EMBO Journal* **24**, 1256–1266.

291. McDonald, H. B., Stewart, R. J., and Goldstein, L. S. (1990). The kinesin-like ncd protein of Drosophila is a minus end-directed microtubule motor. *Cell* **63**, 1159–1165.

292. Meluh, P. B., and Rose, M. D. (1990). KAR3, a kinesin-related gene required for yeast nuclear fusion. *Cell* **60**, 1029–1041.

293. Furuta, K., Edamatsu, M., Maeda, Y., and Toyoshima, Y. Y. (2008). Diffusion and directed movement: *in vitro* motile properties of fission yeast kinesin-14 Pkl1. *The Journal of Biological Chemistry* **283**, 36465–36473.

294. Hirose, K., Loewe, J., Alonso, M., Cross, R. A., and Amos, L. A. (1999). Congruent docking of dimeric kinesin and ncd into three-dimensional electron cryomicroscopy maps of microtubule-motor ADP complexes. *Molecular Biology of the Cell* **10**, 2063–2074.

295. Hallen, M. A., Liang, Z.-Y., and Endow, S. A. (2011). Two-state displacement by the kinesin-14 Ncd stalk. *Biophysical Chemistry* **154**, 56–65.

296. Butterfield, A. E., Stewart, R. J., Schmidt, C. F., and Skliar, M. (2010). Bidirectional power stroke by Ncd kinesin. *Biophysical Journal* **99**, 3905–3915.

297. Bananis, E., Murray, J. W., Stockert, R. J., Satir, P., and Wolkoff, A. W. (2003). Regulation of early endocytic vesicle motility and fission in a reconstituted system. *Journal of Cell Science* **116**, 2749–2761.

298. Reddy, A. S., and Day, I. S. (2001). Kinesins in the Arabidopsis genome: a comparative analysis among eukaryotes. *BMC Genomics* **2**, 2.

299. King, S. M. (2002). Dyneins motor on in plants. *Traffic* **3**, 930–931.

300. Lawrence, C. J., Morris, N. R., Meagher, R. B., and Dawe, R. K. (2001). Dyneins have run their course in plant lineage. *Traffic* **2**, 362–363.

301. Kozielski, F., Sack, S., Marx, A., Thormahlen, M., Schonbrunn, E., Biou, V., Thompson, A., Mandelkow, E. M., and Mandelkow, E. (1997). The crystal structure of dimeric kinesin and implications for microtubule-dependent motility. *Cell* **91**, 985–994.

302. Woehlke, G., and Schliwa, M. (2000). Walking on two heads: the many talents of kinesin. *Nature Reviews Molecular Cell Biology* **1**, 50–58.

303. Scholey, J. M., Heuser, J., Yang, J. T., and Goldstein, L. S. (1989). Identification of globular mechanochemical heads of kinesin. *Nature* **338**, 355–357.

304. Sablin, E. P., Case, R. B., Dai, S. C., Hart, C. L., Ruby, A., Vale, R. D., and Fletterick, R. J. (1998). Direction determination in the minus-end-directed kinesin motor ncd. *Nature* **395**, 813–816.

305. Bloom, G. S., Wagner, M. C., Pfister, K. K., and Brady, S. T. (1988). Native structure and physical properties of bovine brain kinesin and identification of the ATP-binding subunit polypeptide. *Biochemistry* **27**, 3409–3416.

306. Dietrich, K. A., Sindelar, C. V., Brewer, P. D., Downing, K. H., Cremo, C. R., and Rice, S. E. (2008). The kinesin-1 motor protein is regulated by a direct interaction of its head and tail. *Proceedings of the National Academy of Sciences of the United States of America* **105**, 8938–8943.

307. Hackney, D. D., and Stock, M. F. (2008). Kinesin tail domains and Mg^{2+} directly inhibit release of ADP from head domains in the absence of microtubules. *Biochemistry* **47**, 7770–7778.

308. Blasius, T. L., Cai, D., Jih, G. T., Toret, C. P., and Verhey, K. J. (2007). Two binding partners cooperate to activate the molecular motor Kinesin-1. *The Journal of Cell Biology* **176**, 11–17.

309. Verhey, K. J., and Hammond, J. W. (2009). Traffic control: regulation of kinesin motors. *Nature Reviews Molecular Cell Biology* **10**, 765–777.

310. Howard, J., Hudspeth, A. J., and Vale, R. D. (1989). Movement of microtubules by single kinesin molecules. *Nature* **342**, 154–158.

311. Svoboda, K., Schmidt, C. F., Schnapp, B. J., and Block, S. M. (1993). Direct observation of kinesin stepping by optical trapping interferometry. *Nature* **365**, 721–727.

312. Schnitzer, M. J., and Block, S. M. (1997). Kinesin hydrolyses one ATP per 8-nm step. *Nature* **388**, 386–390.

313. Hua, W., Young, E. C., Fleming, M. L., and Gelles, J. (1997). Coupling of kinesin steps to ATP hydrolysis. *Nature* **388**, 390–393.

314. Scholz, T., Vicary, J. A., Jeppesen, G. M., Ulcinas, A., Hörber, J. K., and Antognozzi, M. (2011). Processive behaviour of kinesin observed using micro-fabricated cantilevers. *Nanotechnology* **22**, 095707.

315. Spudich, J. A., and Rock, R. S. (2002). A crossbridge too far. *Nature Cell Biology* **4**, E8–10.

316. Hua, W., Chung, J., and Gelles, J. (2002). Distinguishing inchworm and hand-over-hand processive kinesin movement by neck rotation measurements. *Science* **295**, 844–848.

317. Yildiz, A., and Selvin, P. R. (2005). Kinesin: walking, crawling or sliding along? *Trends in Cell Biology* **15**, 112–120.

318. Kaseda, K., Higuchi, H., and Hirose, K. (2003). Alternate fast and slow stepping of a heterodimeric kinesin molecule. *Nature Cell Biology* **5**, 1079–1082.

319. Asbury, C. L., Fehr, A. N., and Block, S. M. (2003). Kinesin moves by an asymmetric hand-over-hand mechanism. *Science* **302**, 2130–2134.

320. Yildiz, A., Tomishige, M., Vale, R. D., and Selvin, P. R. (2004). Kinesin walks hand-over-hand. *Science* **303**, 676–678.

321. Naber, N., Larson, A., Rice, S., Cooke, R., and Pate, E. (2011). Multiple conformations of the nucleotide site of kinesin family motors in the triphosphate state. *Journal of Molecular Biology* **408**, 628–642.

322. Schaap, Iwan A. T., Carrasco, C., de Pablo, Pedro J., and Schmidt, Christoph F. (2011). Kinesin walks the line: single motors observed by atomic force microscopy. *Biophysical Journal* **100**, 2450–2456.

323. Hoenger, A., Thormahlen, M., Diaz-Avalos, R., Doerhoefer, M., Goldie, K. N., Muller, J., and Mandelkow, E. (2000). A new look at the microtubule binding patterns of dimeric kinesins. *Journal of Molecular Biology* **297**, 1087–1103.

324. Martin, D. S., Fathi, R., Mitchison, T. J., and Gelles, J. (2010). FRET measurements of kinesin neck orientation reveal a structural basis for processivity and asymmetry. *Proceedings of the National Academy of Sciences of the United States of America* **107**, 5453–5458.

325. Ali, M. Y., Krementsova, E. B., Kennedy, G. G., Mahaffy, R., Pollard, T. D., Trybus, K. M., and Warshaw, D. M. (2007). Myosin Va maneuvers through actin intersections and diffuses along microtubules. *Proceedings of the National Academy of Sciences of the United States of America* **104**, 4332–4336.

326. Baker, J. P., and Titus, M. A. (1997). A family of unconventional myosins from the nematode *Caenorhabditis elegans*. *Journal of Molecular Biology* **272**, 523–535.

327. Rayment, I., Rypniewski, W. R., Schmidt-Base, K., Smith, R., Tomchick, D. R., Benning, M. M., Winkelmann, D. A., Wesenberg, G., and Holden, H. M. (1993). Three-dimensional structure of myosin subfragment-1: A molecular motor. *Science* **261**, 50–58.

328. Dominguez, R., Freyzon, Y., Trybus, K. M., and Cohen, C. (1998). Crystal structure of a vertebrate smooth muscle myosin motor domain and its complex with the essential light chain: visualization of the pre-power stroke state. *Cell* **94**, 559–571.

329. Houdusse, A., Kalabokis, V. N., Himmel, D., Szent-Gyorgyi, A. G., and Cohen, C. (1999). Atomic structure of scallop myosin subfragment S1 complexed with MgADP: a novel conformation of the myosin head. *Cell* **97**, 459–470.

330. Cheney, R. E., O'Shea, M. K., Heuser, J. E., Coelho, M. V., Wolenski, J. S., Espreafico, E. M., Forscher, P., Larson, R. E., and Mooseker, M. S. (1993). Brain myosin-V is a two-headed unconventional myosin with motor activity. *Cell* **75**, 13–23.

331. Gillespie, P. G., Albanesi, J. P., Bahler, M., Bement, W. M., Berg, J. S., Burgess, D. R., Burnside, B., Cheney, R. E., Corey, D. P., Coudrier, E., de Lanerolle, P., Hammer, J. A., Hasson, T., Holt, J. R., Hudspeth, A. J., Ikebe, M., Kendrick-Jones, J., Korn, E. D., Li, R., Mercer, J. A., Milligan, R. A., Mooseker, M. S., Ostap, E. M., Petit, C., Pollard, T. D., Sellers, J. R., Soldati, T., and Titus, M. A. (2001). Myosin-I nomenclature. *The Journal of Cell Biology* **155**, 703–704.

332. Foth, B. J., Goedecke, M. C., and Soldati, D. (2006). New insights into myosin evolution and classification. *Proceedings of the National Academy of Sciences of the United States of America* **103**, 3681–3686.

333. Berg, J. S., Powell, B. C., and Cheney, R. E. (2001). A millennial myosin census. *Molecular Biology of the Cell* **12**, 780–794.

334. Coluccio, L. M. (1997). Myosin I. *American Journal of Physiology* **273**, C347–C359.

335. Barylko, B., Binns, D. D., and Albanesi, J. P. (2000). Regulation of the enzymatic and motor activities of myosin I. *Biochimica et Biophysica Acta* **1496**, 23–35.

336. Mooseker, M. S., and Cheney, R. E. (1995). Unconventional myosins. *Annual Review of Cell and Developmental Biology* **11**, 633–675.

337. Pollard, T. D., Doberstein, S. K., and Zot, H. G. (1991). Myosin-I. *Annual Review of Physiology* **53**, 653–681.

338. Wu, X., Jung, G., and Hammer, J. A., 3rd. (2000). Functions of unconventional myosins. *Current Opinion in Cell Biology* **12**, 42–51.

339. Jung, G., Wu, X., and Hammer, J. A., 3rd. (1996). Dictyostelium mutants lacking multiple classic myosin I isoforms reveal combinations of shared and distinct functions. *The Journal of Cell Biology* **133**, 305–323.

340. Ruppert, C., Godel, J., Muller, R. T., Kroschewski, R., Reinhard, J., and Bahler, M. (1995). Localization of the rat myosin I molecules myr 1 and myr 2 and *in vivo* targeting of their tail domains. *Journal of Cell Science* **108 (Pt 12)**, 3775–3786.

341. Wagner, M. C., Barylko, B., and Albanesi, J. P. (1992). Tissue distribution and subcellular localization of mammalian myosin I. *The Journal of Cell Biology* **119**, 163–170.

342. Garcia, J. A., Yee, A. G., Gillespie, P. G., and Corey, D. P. (1998). Localization of myosin-Ibeta near both ends of tip links in frog saccular hair cells. *The Journal of Neuroscience* **18**, 8637–8647.

343. Bahler, M., Kroschewski, R., Stoffler, H. E., and Behrmann, T. (1994). Rat myr 4 defines a novel subclass of myosin I: identification, distribution, localization, and mapping of calmodulin-binding sites with differential calcium sensitivity. *The Journal of Cell Biology* **126**, 375–389.

344. Gillespie, P. G., Wagner, M. C., and Hudspeth, A. J. (1993). Identification of a 120 kd hair-bundle myosin located near stereociliary tips. *Neuron* **11**, 581–594.

345. Raposo, G., Cordonnier, M. N., Tenza, D., Menichi, B., Durrbach, A., Louvard, D., and Coudrier, E. (1999). Association of myosin I alpha with endosomes and lysosomes in mammalian cells. *Molecular Biology of the Cell* **10**, 1477–1494.

346. Cooke, R. (1997). Actomyosin interaction in striated muscle. *Physiological Reviews* **77**, 671–697.

347. Holmes, K. C. (1997). The swinging lever-arm hypothesis of muscle contraction. *Current Biology* **7**, R112–R118.

348. Tan, J. L., Ravid, S., and Spudich, J. A. (1992). Control of nonmuscle myosins by phosphorylation. *Annual Review of Biochemistry* **61**, 721–759.

349. Bresnick, A. R. (1999). Molecular mechanisms of nonmuscle myosin-II regulation. *Current Opinion in Cell Biology* **11**, 26–33.

350. Farah, C. S., and Reinach, F. C. (1995). The troponin complex and regulation of muscle contraction. *The FASEB Journal* **9**, 755–767.

351. Xie, X., Harrison, D. H., Schlichting, I., Sweet, R. M., Kalabokis, V. N., Szent-Gyorgyi, A. G., and Cohen, C. (1994). Structure of the regulatory domain of scallop myosin at 2.8 A resolution. *Nature* **368**, 306–312.

352. Sellers, J. R. (2000). Myosins: a diverse superfamily. *Biochimica et Biophysica Acta* **1496**, 3–22.

353. Montell, C., and Rubin, G. M. (1988). The *Drosophila* ninaC locus encodes two photoreceptor cell specific proteins with domains homologous to protein kinases and the myosin heavy chain head. *Cell* **52**, 757–772.

354. Battelle, B. A., Andrews, A. W., Calman, B. G., Sellers, J. R., Greenberg, R. M., and Smith, W. C. (1998). A myosin III from Limulus eyes is a clock-regulated phosphoprotein. *The Journal of Neuroscience* **18**, 4548–4559.

355. Montell, C. (1999). Visual transduction in *Drosophila. Annual Review of Cell and Developmental Biology* **15**, 231–268.

356. Wes, P. D., Xu, X. Z., Li, H. S., Chien, F., Doberstein, S. K., and Montell, C. (1999). Termination of phototransduction requires binding of the NINAC myosin III and the PDZ protein INAD. *Nature Neuroscience* **2**, 447–453.

357. Bahler, M. (2000). Are class III and class IX myosins motorized signalling molecules? *Biochimica et Biophysica Acta* **1496**, 52–59.

358. Weber, K. L., Sokac, A. M., Berg, J. S., Cheney, R. E., and Bement, W. M. (2004). A microtubule-binding myosin required for nuclear anchoring and spindle assembly. *Nature* **431**, 325–329.

359. Titus, M. A. (1997). Motor proteins: myosin V: the multi-purpose transport motor. *Current Biology* **7**, R301–R304.

360. Reck-Peterson, S. L., Provance, D. W., Jr., Mooseker, M. S., and Mercer, J. A. (2000). Class V myosins. *Biochimica et Biophysica Acta* **1496**, 36–51.

361. Bement, W. M., Hasson, T., Wirth, J. A., Cheney, R. E., and Mooseker, M. S. (1994). Identification and overlapping expression of multiple unconventional myosin genes in vertebrate cell types. *Proceedings of the National Academy of Sciences of the United States of America* **91**, 6549–6553.

362. Brown, S. S. (1997). Myosins in yeast. *Current Opinion in Cell Biology* **9**, 44–48.

363. Espindola, F. S., Espreafico, E. M., Coelho, M. V., Martins, A. R., Costa, F. R., Mooseker, M. S., and Larson, R. E. (1992). Biochemical and immunological characterization of p190-calmodulin complex from vertebrate brain: a novel calmodulin-binding myosin. *The Journal of Cell Biology* **118**, 359–368.

364. Espreafico, E. M., Cheney, R. E., Matteoli, M., Nascimento, A. A., De Camilli, P. V., Larson, R. E., and Mooseker, M. S. (1992). Primary structure and cellular localization of chicken brain myosin-V (p190), an unconventional myosin with calmodulin light chains. *The Journal of Cell Biology* **119**, 1541–1557.

365. Ponting, C. P. (1995). AF-6/cno: neither a kinesin nor a myosin, but a bit of both. *Trends in Biochemical Sciences* **20**, 265–266.

366. Moore, J. R., Krementsova, E. B., Trybus, K. M., and Warshaw, D. M. (2001). Myosin V exhibits a high duty cycle and large unitary displacement. *The Journal of Cell Biology* **155**, 625–635.

367. Harris, D. E., and Warshaw, D. M. (1993). Smooth and skeletal muscle myosin both exhibit low duty cycles at zero load *in vitro*. *The Journal of Biological Chemistry* **268**, 14764–14768.

368. Rief, M., Rock, R. S., Mehta, A. D., Mooseker, M. S., Cheney, R. E., and Spudich, J. A. (2000). Myosin-V stepping kinetics: a molecular model for processivity. *Proceedings of the National Academy of Sciences of the United States of America* **97**, 9482–9486.

369. Veigel, C., Wang, F., Bartoo, M. L., Sellers, J. R., and Molloy, J. E. (2002). The gated gait of the processive molecular motor, myosin V. *Nature Cell Biology* **4**, 59–65.

370. Walker, M. L., Burgess, S. A., Sellers, J. R., Wang, F., Hammer, J. A., 3rd, Trinick, J., and Knight, P. J. (2000). Two-headed binding of a processive myosin to F-actin. *Nature* **405**, 804–807.

371. Purcell, T. J., Naber, N., Sutton, S., Cooke, R., and Pate, E. (2011). EPR spectra and molecular dynamics agree that the nucleotide pocket of myosin v is closed and that it opens on binding actin. *Journal of Molecular Biology* **411**, 16–26.

372. Wu, X., Bowers, B., Wei, Q., Kocher, B., and Hammer, J. A., 3rd. (1997). Myosin V associates with melanosomes in mouse melanocytes: evidence that myosin V is an organelle motor. *Journal of Cell Science* **110(Pt 7)**, 847–859.

373. Tabb, J. S., Molyneaux, B. J., Cohen, D. L., Kuznetsov, S. A., and Langford, G. M. (1998). Transport of ER vesicles on actin filaments in neurons by myosin V. *Journal of Cell Science* **111(Pt 21)**, 3221–3234.

374. Rodionov, V. I., Hope, A. J., Svitkina, T. M., and Borisy, G. G. (1998). Functional coordination of microtubule-based and actin-based motility in melanophores. *Current Biology* **8**, 165–168.

375. Arn, E. A., and Macdonald, P. M. (1998). Motors driving mRNA localization: new insights from *in vivo* imaging. *Cell* **95**, 151–154.

376. Kellerman, K. A., and Miller, K. G. (1992). An unconventional myosin heavy chain gene from *Drosophila* melanogaster. *The Journal of Cell Biology* **119**, 823–834.

377. Wells, A. L., Lin, A. W., Chen, L. Q., Safer, D., Cain, S. M., Hasson, T., Carragher, B. O., Milligan, R. A., and Sweeney, H. L. (1999). Myosin VI is an actin-based motor that moves backwards. *Nature* **401**, 505–508.

378. Song, C. F., Sader, K., White, H., Kendrick-Jones, J., and Trinick, J. (2010). Nucleotide-dependent shape changes in the reverse direction motor, myosin VI. *Biophysical Journal* **99**, 3336–3344.

379. Chuan, P., Spudich, J. A., and Dunn, A. R. (2011). Robust Mechanosensing and Tension Generation by Myosin VI. *Journal of Molecular Biology* **405**, 105–112.

380. Mermall, V., McNally, J. G., and Miller, K. G. (1994). Transport of cytoplasmic particles catalysed by an unconventional myosin in living *Drosophila* embryos. *Nature* **369**, 560–562.

381. Lantz, V. A., and Miller, K. G. (1998). A class VI unconventional myosin is associated with a homologue of a microtubule-binding protein, cytoplasmic linker protein-170, in neurons and at the posterior pole of *Drosophila* embryos. *The Journal of Cell Biology* **140**, 897–910.

382. Buss, F., Kendrick-Jones, J., Lionne, C., Knight, A. E., Cote, G. P., and Paul Luzio, J. (1998). The localization of myosin VI at the golgi complex and leading edge of fibroblasts and its phosphorylation and recruitment into membrane ruffles of A431 cells after growth factor stimulation. *The Journal of Cell Biology* **143**, 1535–1545.

383. Hasson, T. (1997). Unconventional myosins, the basis for deafness in mouse and man. *The American Journal of Human Genetics* **61**, 801–805.

384. Avraham, K. B., Hasson, T., Steel, K. P., Kingsley, D. M., Russell, L. B., Mooseker, M. S., Copeland, N. G., and Jenkins, N. A. (1995). The mouse Snell's waltzer deafness gene encodes an unconventional myosin required for structural integrity of inner ear hair cells. *Nature Genetics* **11**, 369–375.

385. Yang, Y., Kovacs, M., Sakamoto, T., Zhang, F., Kiehart, D. P., and Sellers, J. R. (2006). Dimerized *Drosophila* myosin VIIa: a processive motor. *Proceedings of the National Academy of Sciences of the United States of America* **103**, 5746–5751.

386. Haithcock, J., Billington, N., Choi, K., Fordham, J., Sellers, J. R., Stafford, W. F., White, H., and Forgacs, E. (2011). The kinetic mechanism of mouse myosin VIIA. *The Journal of Biological Chemistry* **286**, 8819–8828.

387. Titus, M. A. (1999). A class VII unconventional myosin is required for phagocytosis. *Current Biology* **9**, 1297–1303.

388. Tuxworth, R. I., Weber, I., Wessels, D., Addicks, G. C., Soll, D. R., Gerisch, G., and Titus, M. A. (2001). A role for myosin VII in dynamic cell adhesion. *Current Biology* **11**, 318–329.

389. Knight, A. E., and Kendrick-Jones, J. (1993). A myosin-like protein from a higher plant. *Journal of Molecular Biology* **231**, 148–154.

390. Baluska, F., Cvrckova, F., Kendrick-Jones, J., and Volkmann, D. (2001). Sink plasmodesmata as gateways for phloem unloading. Myosin VIII and calreticulin as molecular determinants of sink strength? *Plant Physiology* **126**, 39–46.

391. Reichelt, S., Knight, A. E., Hodge, T. P., Baluska, F., Samaj, J., Volkmann, D., and Kendrick-Jones, J. (1999). Characterization of the unconventional myosin VIII in plant cells and its localization at the post-cytokinetic cell wall. *The Plant Journal* **19**, 555–567.

392. Golomb, L., Abu-Abied, M., Belausov, E., and Sadot, E. (2008). Different subcellular localizations and functions of Arabidopsis myosin VIII. *BMC Plant Biology* **8**, 3.

393. Chieregatti, E., Gartner, A., Stoffler, H. E., and Bahler, M. (1998). Myr 7 is a novel myosin IX-RhoGAP expressed in rat brain. *Journal of Cell Science* **111(Pt 24)**, 3597–3608.

394. Gorman, S. W., Haider, N. B., Grieshammer, U., Swiderski, R. E., Kim, E., Welch, J. W., Searby, C., Leng, S., Carmi, R., Sheffield, V. C., and Duhl, D. M. (1999). The cloning and developmental expression of unconventional myosin IXA (MYO9A) a gene in the Bardet-Biedl syndrome (BBS4) region at chromosome 15q22-q23. *Genomics* **59**, 150–160.

395. Post, P. L., Bokoch, G. M., and Mooseker, M. S. (1998). Human myosin-IXb is a mechanochemically active motor and a GAP for rho. *Journal of Cell Science* **111(Pt 7)**, 941–950.

396. Reinhard, J., Scheel, A. A., Diekmann, D., Hall, A., Ruppert, C., and Bahler, M. (1995). A novel type of myosin implicated in signalling by rho family GTPases. *The EMBO Journal* **14**, 697–704.

397. Post, P. L., Tyska, M. J., O'Connell, C. B., Johung, K., Hayward, A., and Mooseker, M. S. (2002). Myosin-IXb is a single-headed and processive motor. *The Journal of Biological Chemistry* **277**, 11679–11683.

398. Inoue, A., Saito, J., Ikebe, R., and Ikebe, M. (2002). Myosin IXb is a single-headed minus-end-directed processive motor. *Nature Cell Biology* **4**, 302–306.

399. Solc, C. K., Derfler, B. H., Duyk, G. M., and Corey, D. P. (1994). PCR screen of a Bull-frog inner ear library reveals a novel class of myosins (class X). *Auditory Neuroscience* **1**, 63–75.

400. Berg, J. S., and Cheney, R. E. (2002). Myosin-X is an unconventional myosin that undergoes intrafilopodial motility. *Nature Cell Biology* **4**, 246–250.

401. Sousa, A. D., and Cheney, R. E. (2005). Myosin-X: a molecular motor at the cell's fingertips. *Trends in Cell Biology* **15**, 533–539.

402. Li, J. F., and Nebenfuhr, A. (2007). Organelle targeting of myosin XI is mediated by two globular tail subdomains with separate cargo binding sites. *The Journal of Biological Chemistry* **282**, 20593–20602.

403. Kinkema, M., and Schiefelbein, J. (1994). A myosin from a higher plant has structural similarities to class V myosins. *Journal of Molecular Biology* **239**, 591–597.

404. Tominaga, M., Kojima, H., Yokota, E., Orii, H., Nakamori, R., Katayama, E., Anson, M., Shimmen, T., and Oiwa, K. (2003). Higher plant myosin XI moves processively on actin with 35 nm steps at high velocity. *The EMBO Journal* **22**, 1263–1272.

405. Higashi-Fujime, S., Ishikawa, R., Iwasawa, H., Kagami, O., Kurimoto, E., Kohama, K., and Hozumi, T. (1995). The fastest actin-based motor protein from the green algae, Chara, and its distinct mode of interaction with actin. *FEBS Letters* **375**, 151–154.

406. Morimatsu, M., Nakamura, A., Sumiyoshi, H., Sakaba, N., Taniguchi, H., Kohama, K., and Higashi-Fujime, S. (2000). The molecular structure of the fastest myosin from green algae, Chara. *Biochemical and Biophysical Research Communications* **270**, 147–152.

407. Shimmen, T., and Yokota, E. (2004). Cytoplasmic streaming in plants. *Current Opinion in Cell Biology* **16**, 68–72.

408. Vidali, L., Burkart, G. M., Augustine, R. C., Kerdavid, E., Tuzel, E., and Bezanilla, M. (2010). Myosin XI Is Essential for Tip Growth in Physcomitrella patens. *Plant Cell* **22(6)**, 1868–1882.

409. Sparkes, I. A., Teanby, N. A., and Hawes, C. (2008). Truncated myosin XI tail fusions inhibit peroxisome, Golgi, and mitochondrial movement in tobacco leaf epidermal cells: a genetic tool for the next generation. *Journal of Experimental Botany* **59**, 2499–2512.

410. Vugrek, O., Sawitzky, H., and Menzel, D. (2003). Class XIII myosins from the green alga Acetabularia: driving force in organelle transport and tip growth? *Journal of Muscle Research and Cell Motility* **24**, 87–97.

411. Heintzelman, M. B., and Schwartzman, J. D. (1997). A novel class of unconventional myosins from Toxoplasma gondii. *Journal of Molecular Biology* **271**, 139–146.

412. Hettmann, C., Herm, A., Geiter, A., Frank, B., Schwarz, E., Soldati, T., and Soldati, D. (2000). A dibasic motif in the tail of a class XIV apicomplexan myosin is an essential determinant of plasma membrane localization. *Molecular Biology of the Cell* **11**, 1385–1400.

413. Liang, Y., Wang, A., Belyantseva, I. A., Anderson, D. W., Probst, F. J., Barber, T. D., Miller, W., Touchman, J. W., Jin, L., Sullivan, S. L., Sellers, J. R., Camper, S. A., Lloyd, R. V., Kachar, B., Friedman, T. B., and Fridell, R. A. (1999). Characterization of the human and mouse unconventional myosin XV genes responsible for hereditary deafness DFNB3 and shaker 2. *Genomics* **61**, 243–258.

414. Wang, A., Liang, Y., Fridell, R. A., Probst, F. J., Wilcox, E. R., Touchman, J. W., Morton, C. C., Morell, R. J., Noben-Trauth, K., Camper, S. A., and Friedman, T. B. (1998). Association of unconventional myosin MYO15 mutations with human nonsyndromic deafness DFNB3. *Science* **280**, 1447–1451.

415. Probst, F. J., Fridell, R. A., Raphael, Y., Saunders, T. L., Wang, A., Liang, Y., Morell, R. J., Touchman, J. W., Lyons, R. H., Noben-Trauth, K., Friedman, T. B., and Camper, S. A. (1998). Correction of deafness in shaker-2 mice by an unconventional myosin in a BAC transgene. *Science* **280**, 1444–1447.

416. Tzolovsky, G., Millo, H., Pathirana, S., Wood, T., and Bownes, M. (2002). Identification and phylogenetic analysis of *Drosophila* melanogaster myosins. *Molecular Biology and Evolution* **19**, 1041–1052.

417. Jacinto, A., Woolner, S., and Martin, P. (2002). Dynamic analysis of dorsal closure in *Drosophila*: from genetics to cell biology. *Developmental Cell* **3**, 9–19.

418. Liu, R., Woolner, S., Johndrow, J. E., Metzger, D., Flores, A., and Parkhurst, S. M. (2008). Sisyphus, the *Drosophila* myosin XV homolog, traffics within filopodia transporting key sensory and adhesion cargos. *Development* **135**, 53–63.

419. Patel, K. G., Liu, C., Cameron, P. L., and Cameron, R. S. (2001). Myr 8, a novel unconventional myosin expressed during brain development associates with the protein phosphatase catalytic subunits 1alpha and 1gamma1. *The Journal of Neuroscience* **21**, 7954–7968.

420. Richards, T. A., and Cavalier-Smith, T. (2005). Myosin domain evolution and the primary divergence of eukaryotes. *Nature* **436**, 1113–1118.

421. Cameron, R. S., Liu, C., Mixon, A. S., Pihkala, J. P., Rahn, R. J., and Cameron, P. L. (2007). Myosin16b: The COOH-tail region directs localization to the nucleus and overexpression delays S-phase progression. *Cell Motility and the Cytoskeleton* **64**, 19–48.

422. Hodge, T., and Cope, M. J. (2000). A myosin family tree. *Journal of Cell Science* **113(Pt 19)**, 3353–3354.

423. Takeshita, N., Ohta, A., and Horiuchi, H. (2005). CsmA, a class V chitin synthase with a myosin motor-like domain, is localized through direct interaction with the actin cytoskeleton in Aspergillus nidulans. *Molecular Biology of the Cell* **16**, 1961–1970.

424. Werner, S., Sugui, J. A., Steinberg, G., and Deising, H. B. (2007). A chitin synthase with a myosin-like motor domain is essential for hyphal growth, appressorium differentiation, and pathogenicity of the maize anthracnose fungus Colletotrichum graminicola. *Molecular Plant-Microbe Interactions* **20**, 1555–1567.

425. Furusawa, T., Ikawa, S., Yanai, N., and Obinata, M. (2000). Isolation of a novel PDZ-containing myosin from hematopoietic supportive bone marrow stromal cell lines. *Biochemical and Biophysical Research Communications* **270**, 67–75.

426. Mori, K., Furusawa, T., Okubo, T., Inoue, T., Ikawa, S., Yanai, N., Mori, K. J., and Obinata, M. (2003). Genome structure and differential expression of two isoforms of a novel PDZ-containing myosin (MysPDZ) (Myo18A). *Journal of Biochemistry* **133**, 405–413.

427. Yamashita, R. A., Sellers, J. R., and Anderson, J. B. (2000). Identification and analysis of the myosin superfamily in Drosophila: a database approach. *Journal of Muscle Research and Cell Motility* **21**, 491–505.

428. Ranganathan, R., and Ross, E. M. (1997). PDZ domain proteins: Scaffolds for signaling complexes. *Current Biology* **7**, R770–R773.

429. Isogawa, Y., Kon, T., Inoue, T., Ohkura, R., Yamakawa, H., Ohara, O., and Sutoh, K. (2005). The N-terminal domain of MYO18A has an ATP-insensitive actin-binding site. *Biochemistry* **44**, 6190–6196.

430. Salamon, M., Millino, C., Raffaello, A., Mongillo, M., Sandri, C., Bean, C., Negrisolo, E., Pallavicini, A., Valle, G., Zaccolo, M., Schiaffino, S., and Lanfranchi, G. (2003). Human MYO18B, a novel unconventional myosin heavy chain expressed in striated muscles moves into the myonuclei upon differentiation. *Journal of Molecular Biology* **326**, 137–149.

431. Quintero, O. A., DiVito, M. M., Adikes, R. C., Kortan, M. B., Case, L. B., Lier, A. J., Panaretos, N. S., Slater, S. Q., Rengarajan, M., Feliu, M., and Cheney, R. E. (2009). Human Myo19 is a novel myosin that associates with mitochondria. *Current Biology* **19**, 2008–2013.

432. Lemmon, M. A. (2003). Phosphoinositide recognition domains. *Traffic* **4**, 201–213.

433. Macias, M. J., Wiesner, S., and Sudol, M. (2002). WW and SH3 domains, two different scaffolds to recognize proline-rich ligands. *FEBS Letters* **513**, 30–37.

434. Buchberger, A. (2002). From UBA to UBX: New words in the ubiquitin vocabulary. *Trends in Cell Biology* **12**, 216–221.

435. Vale, R. D. (2003). Myosin V motor proteins: marching stepwise towards a mechanism. *The Journal of Cell Biology* **163**, 445–450.

436. Hudson, A. M., and Cooley, L. (2008). Phylogenetic, structural and functional relationships between WD- and Kelch-repeat proteins. *Subcellular Biochemistry* **48**, 6–19.

437. Smith, C. A., and Rayment, I. (1996). X-ray structure of the magnesium(II). ADP.vanadate complex of the Dictyostelium discoideum myosin motor domain to 1.9 A resolution. *Biochemistry* **35**, 5404–5417.

438. Minoda, H., Okabe, T., Inayoshi, Y., Miyakawa, T., Miyauchi, Y., Tanokura, M., Katayama, E., Wakabayashi, T., Akimoto, T., and Sugi, H. (2011). Electron microscopic evidence for the myosin head lever arm mechanism in hydrated myosin filaments using the gas environmental chamber. *Biochemical and Biophysical Research Communications* **405**, 651–656.

439. Sun, Y., and Goldman, Yale E. (2011). Lever-arm mechanics of processive Myosins. *Biophysical Journal* **101**, 1–11.

440. Geeves, M. A., and Holmes, K. C. (1999). Structural mechanism of muscle contraction. *Annual Review of Biochemistry* **68**, 687–728.

441. Brown, J. H., Senthil Kumar, V. S., O'Neall-Hennessey, E., Reshetnikova, L., Robinson, H., Nguyen-McCarty, M., Szent-Györgyi, A. G., and Cohen, C. (2011). Visualizing key hinges and a potential major source of compliance in the lever arm of myosin. *Proceedings of the National Academy of Sciences of the United States of America* **108**, 114–119.

442. Nesmelov, Y. E., Agafonov, R. V., Negrashov, I. V., Blakely, S. E., Titus, M. A., and Thomas, D. D. (2011). Structural kinetics of myosin by transient time-resolved FRET. *Proceedings of the National Academy of Sciences of the United States of America* **108**, 1891–1896.

443. Coureux, P. D., Sweeney, H. L., and Houdusse, A. (2004). Three myosin V structures delineate essential features of chemo-mechanical transduction. *The EMBO Journal* **23**, 4527–4537.

444. Hodges, A. R., Krementsova, E. B., and Trybus, K. M. (2007). Engineering the processive run length of Myosin V. *The Journal of Biological Chemistry* **282**, 27192–27197.

445. Li, X. D., Jung, H. S., Mabuchi, K., Craig, R., and Ikebe, M. (2006). The globular tail domain of myosin Va functions as an inhibitor of the myosin Va motor. *The Journal of Biological Chemistry* **281**, 21789–21798.

446. Cross, R. (2010). Kinesin-14: the roots of reversal. *BMC Biology* **8**, 107.

447. Wickstead, B., and Gull, K. (2007). Dyneins across eukaryotes: A comparative genomic analysis. *Traffic* **8**, 1708–1721.

448. Gibbons, I. R. (1995). Dynein family of motor proteins: present status and future questions. *Cell Motility and the Cytoskeleton* **32**, 136–144.

449. Toba, S., Watanabe, T. M., Yamaguchi-Okimoto, L., Toyoshima, Y. Y., and Higuchi, H. (2006). Overlapping hand-over-hand mechanism of single molecular motility of cytoplasmic dynein. *Proceedings of the National Academy of Sciences of the United States of America* **103**, 5741–5745.

450. Mallik, R., Carter, B. C., Lex, S. A., King, S. J., and Gross, S. P. (2004). Cytoplasmic dynein functions as a gear in response to load. *Nature* **427**, 649–652.

451. Vallee, R. B., Williams, J. C., Varma, D., and Barnhart, L. E. (2004). Dynein: An ancient motor protein involved in multiple modes of transport. *Journal of Neurobiology* **58**, 189–200.

452. Karki, S., and Holzbaur, E. L. (1999). Cytoplasmic dynein and dynactin in cell division and intracellular transport. *Current Opinion in Cell Biology* **11**, 45–53.

453. Hook, P., and Vallee, R. B. (2006). The dynein family at a glance. *Journal of Cell Science* **119**, 4369–4371.

454. King, S. M. (2000). The dynein microtubule motor. *Biochimica et Biophysica Acta* **1496**, 60–75.

455. Pfister, K. K., Fisher, E. M. C., Gibbons, I. R., Hays, T. S., Holzbaur, E. L. F., McIntosh, J. R., Porter, M. E., Schroer, T. A., Vaughan, K. T., Witman, G. B., King, S. M., and Vallee, R. B. (2005). Cytoplasmic dynein nomenclature. *The Journal of Cell Biology* **171**, 411–413.

456. King, S. M. (2000). AAA domains and organization of the dynein motor unit. *Journal of Cell Science* **113(Pt 14)**, 2521–2526.

457. Asai, D. J., and Koonce, M. P. (2001). The dynein heavy chain: structure, mechanics and evolution. *Trends in Cell Biology* **11**, 196–202.

458. Gee, M. A., Heuser, J. E., and Vallee, R. B. (1997). An extended microtubule-binding structure within the dynein motor domain. *Nature* **390**, 636–639.

459. Vallee, R. B., and Gee, M. A. (1998). Make room for dynein. *Trends in Cell Biology* **8**, 490–494.

460. Koonce, M. P., and Tikhonenko, I. (2000). Functional elements within the dynein microtubule-binding domain. *Molecular Biology of the Cell* **11**, 523–529.

461. Pfister, K. K., Fisher, E. M., Gibbons, I. R., Hays, T. S., Holzbaur, E. L., McIntosh, J. R., Porter, M. E., Schroer, T. A., Vaughan, K. T., Witman, G. B., King, S. M., and Vallee, R. B. (2005). Cytoplasmic dynein nomenclature. *The Journal of Cell Biology* **171**, 411–413.

462. Luck, D. J. (1984). Genetic and biochemical dissection of the eucaryotic flagellum. *The Journal of Cell Biology* **98**, 789–794.

463. Summers, K. E., and Gibbons, I. R. (1971). Adenosine triphosphate-induced sliding of tubules in trypsin-treated flagella of sea-urchin sperm. *Proceedings of the National Academy of Sciences of the United States of America* **68**, 3092–3096.

464. Lindemann, C. B., and Lesich, K. A. (2010). Flagellar and ciliary beating: the proven and the possible. *Journal of Cell Science* **123**, 519–528.

465. Porter, M. E., and Sale, W. S. (2000). The 9 + 2 axoneme anchors multiple inner arm dyneins and a network of kinases and phosphatases that control motility. *The Journal of Cell Biology* **151**, F37–F42.

466. Goodenough, U. W., and Heuser, J. E. (1985). Substructure of inner dynein arms, radial spokes, and the central pair/projection complex of cilia and flagella. *The Journal of Cell Biology* **100**, 2008–2018.

467. Hopkins, J. M. (1970). Subsidiary components of the flagella of Chlamydomonas reinhardii. *Journal of Cell Science* **7**, 823–839.

468. Piperno, G., Mead, K., and Shestak, W. (1992). The inner dynein arms I2 interact with a "dynein regulatory complex" in Chlamydomonas flagella. *The Journal of Cell Biology* **118**, 1455–1463.

469. Smith, E. F. (2002). Regulation of flagellar dynein by calcium and a role for an axonemal calmodulin and calmodulin-dependent kinase. *Molecular Biology of the Cell* **13**, 3303–3313.

470. Brokaw, C. J. (1971). Bend propagation by a sliding filament model for flagella. *Journal of Experimental Biology* **55**, 289–304.

471. Brokaw, C. J. (1972). Flagellar movement: A sliding filament model. *Science* **178**, 455–462.

472. Brokaw, C. J. (1972). Computer simulation of flagellar movement. I. Demonstration of stable bend propagation and bend initiation by the sliding filament model. *Biophysical Journal* **12**, 564–586.

473. Haimo, L. T., Telzer, B. R., and Rosenbaum, J. L. (1979). Dynein binds to and crossbridges cytoplasmic microtubules. *Proceedings of the National Academy of Sciences of the United States of America* **76**, 5759–5763.

474. Burgess, S. A., Walker, M. L., Sakakibara, H., Knight, P. J., and Oiwa, K. (2003). Dynein structure and power stroke. *Nature* **421**, 715–718.

475. Ichikawa, M., Watanabe, Y., Murayama, T., and Toyoshima, Y. Y. (2011). Recombinant human cytoplasmic dynein heavy chain 1 and 2: Observation of dynein-2 motor activity *in vitro*. *FEBS Letters* **585**, 2419–2423.

476. Koonce, M. P., and Samso, M. (1996). Overexpression of cytoplasmic dynein's globular head causes a collapse of the interphase microtubule network in Dictyostelium. *Molecular Biology of the Cell* **7**, 935–948.

477. Mocz, G., and Gibbons, I. R. (2001). Model for the motor component of dynein heavy chain based on homology to the AAA family of oligomeric ATPases. *Structure* **9**, 93–103.

478. Imamula, K., Kon, T., Ohkura, R., and Sutoh, K. (2007). The coordination of cyclic microtubule association/dissociation and tail swing of cytoplasmic dynein. *Proceedings of the National Academy of Sciences of the United States of America* **104**, 16134–16139.

479. Takahashi, Y., Edamatsu, M., and Toyoshima, Y. Y. (2004). Multiple ATP-hydrolyzing sites that potentially function in cytoplasmic dynein. *Proceedings of the National Academy of Sciences of the United States of America* **101**, 12865–12869.

480. Kon, T., Nishiura, M., Ohkura, R., Toyoshima, Y. Y., and Sutoh, K. (2004). Distinct functions of nucleotide-binding/hydrolysis sites in the four AAA modules of cytoplasmic dynein. *Biochemistry* **43**, 11266–11274.

481. Cho, C., Reck-Peterson, S. L., and Vale, R. D. (2008). Regulatory ATPase sites of cytoplasmic dynein affect processivity and force generation. *The Journal of Biological Chemistry* **283**, 25839–25845.

482. Gibbons, I. R., Garbarino, J. E., Tan, C. E., Reck-Peterson, S. L., Vale, R. D., and Carter, A. P. (2005). The affinity of the dynein microtubule-binding domain is modulated by the conformation of its coiled coil stalk. *The Journal of Biological Chemistry* **280**, 23960–23965.

483. Kon, T., Imamula, K., Roberts, A. J., Ohkura, R., Knight, P. J., Gibbons, I. R., Burgess, S. A., and Sutoh, K. (2009). Helix sliding in the stalk coiled coil of dynein couples ATPase and microtubule binding. *Nature Structural and Molecular Biology* **16**, 325–333.

484. Carter, A. P., and Vale, R. D. (2010). Communication between the AAA^+ ring and microtubule-binding domain of dynein. *Biochemistry and Cell Biology* **88**, 15–21.

485. Reck-Peterson, S. L., Yildiz, A., Carter, A. P., Gennerich, A., Zhang, N., and Vale, R. D. (2006). Single-molecule analysis of dynein processivity and stepping behavior. *Cell* **126**, 335–348.

486. Finkelstein, I. J., Visnapuu, M. L., and Greene, E. C. (2010). Single-molecule imaging reveals mechanisms of protein disruption by a DNA translocase. *Nature* **468**, 983–987.

487. Janscak, P., MacWilliams, M. P., Sandmeier, U., Nagaraja, V., and Bickle, T. A. (1999). DNA translocation blockage, a general mechanism of cleavage site selection by type I restriction enzymes. *European Molecular Biology Organisation Journal* **18**, 2638–2647.

488. Zhao, X., and Jain, C. (2011). DEAD-Box proteins from *Escherichia coli* exhibit multiple ATP-independent activities. *Journal of Bacteriology* **193**, 2236–2241.

489. Erzberger, J. P., and Berger, J. M. (2006). Evolutionary relationships and structural mechanisms of AAA$^+$ proteins. *Annual Review of Biophysics and Biomolecular Structure* **35**, 93–114.

490. Walker, J. E., Saraste, M., Runswick, M. J., and Gay, N. J. (1982). Distantly related sequences in the alpha- and beta-subunits of ATP synthase, myosin, kinases and other ATP-requiring enzymes and a common nucleotide binding fold. *The EMBO Journal* **1**, 945–951.

491. Scheffzek, K., Ahmadian, M. R., Kabsch, W., Wiesmuller, L., Lautwein, A., Schmitz, F., and Wittinghofer, A. (1997). The Ras-RasGAP complex: structural basis for GTPase activation and its loss in oncogenic Ras mutants. *Science* **277**, 333–338.

492. Ali, J. A., Maluf, N. K., and Lohman, T. M. (1999). An oligomeric form of *E. coli* UvrD is required for optimal helicase activity. *Journal of Molecular Biology* **293**, 815–834.

493. Byrd, A. K., and Raney, K. D. (2004). Protein displacement by an assembly of helicase molecules aligned along single-stranded DNA. *Nature Structural and Molecular Biology* **11**, 531–538.

494. Byrd, A. K., and Raney, K. D. (2005). Increasing the length of the single-stranded overhang enhances unwinding of duplex DNA by bacteriophage T4 Dda helicase. *Biochemistry* **44**, 12990–12997.

495. Levin, M. K., Wang, Y. H., and Patel, S. S. (2004). The functional interaction of the hepatitis C virus helicase molecules is responsible for unwinding processivity. *The Journal of Biological Chemistry* **279**, 26005–26012.

496. Caruthers, J. M., and McKay, D. B. (2002). Helicase structure and mechanism. *Current Opinion in Structural Biology* **12**, 123–133.

497. Marintcheva, B., and Weller, S. K. (2003). Helicase motif Ia is involved in single-strand DNA-binding and helicase activities of the herpes simplex virus type 1 origin-binding protein, UL9. *Journal of Virology* **77**, 2477–2488.

498. Pause, A., and Sonenberg, N. (1992). Mutational analysis of a DEAD box RNA helicase: the mammalian translation initiation factor eIF-4A. *The EMBO Journal* **11**, 2643–2654.

499. Korolev, S., Yao, N., Lohman, T. M., Weber, P. C., and Waksman, G. (1998). Comparisons between the structures of HCV and Rep helicases reveal structural similarities between SF1 and SF2 super-families of helicases. *Protein Science* **7**, 605–610.

500. Hall, M. C., and Matson, S. W. (1999). Helicase motifs: The engine that powers DNA unwinding. *Molecular Microbiology* **34**, 867–877.

501. Tomko, E. J., Jia, H., Park, J., Maluf, N. K., Ha, T., and Lohman, T. M. (2010). 5[prime]-Single-stranded/duplex DNA junctions are loading sites for *E. coli* UvrD translocase. *The EMBO Journal* **29**, 3826–3839.

502. Ha, T., Rasnik, I., Cheng, W., Babcock, H. P., Gauss, G. H., Lohman, T. M., and Chu, S. (2002). Initiation and re-initiation of DNA unwinding by the *Escherichia coli* Rep helicase. *Nature* **419**, 638–641.

503. Wong, I., and Lohman, T. M. (1996). ATPase activity of *Escherichia coli* Rep helicase crosslinked to single-stranded DNA: Implications for ATP driven helicase translocation. *Proceedings of the National Academy of Sciences of the United States of America* **93**, 10051–10056.

504. Dittrich, M., and Schulten, K. (2006). PcrA Helicase, a Prototype ATP-Driven Molecular Motor. *Structure* **14**, 1345–1353.

505. Subramanya, H. S., Bird, L. E., Brannigan, J. A., and Wigley, D. B. (1996). Crystal structure of a DExx box DNA helicase. *Nature* **384**, 379–383.

506. Korolev, S., Hsieh, J., Gauss, G. H., Lohman, T. M., and Waksman, G. (1997). Major domain swiveling revealed by the crystal structures of complexes of *E. coli* Rep helicase bound to single-stranded DNA and ADP. *Cell* **90**, 635–647.

507. Velankar, S. S., Soultanas, P., Dillingham, M. S., Subramanya, H. S., and Wigley, D. B. (1999). Crystal structures of complexes of PcrA DNA helicase with a DNA substrate indicate an inchworm mechanism. *Cell* **97**, 75–84.

508. Soultanas, P., Dillingham, M. S., Velankar, S. S., and Wigley, D. B. (1999). DNA binding mediates conformational changes and metal ion coordination in the active site of PcrA helicase. *Journal of Molecular Biology* **290**, 137–148.

509. Soultanas, P., Dillingham, M. S., Wiley, P., Webb, M. R., and Wigley, D. B. (2000). Uncoupling DNA translocation and helicase activity in PcrA: Direct evidence for an active mechanism. *The EMBO Journal* **19**, 3799–3810.

510. Dillingham, M. S., Wigley, D. B., and Webb, M. R. (2000). Demonstration of unidirectional single-stranded DNA translocation by PcrA helicase: Measurement of step size and translocation speed. *Biochemistry* **39**, 205–212.

511. Dillingham, M. S., Soultanas, P., and Wigley, D. B. (1999). Site-directed mutagenesis of motif III in PcrA helicase reveals a role in coupling ATP hydrolysis to strand separation. *Nucleic Acids Research* **27**, 3310–3317.

512. Dillingham, M. S., Soultanas, P., Wiley, P., Webb, M. R., and Wigley, D. B. (2001). Defining the roles of individual residues in the single-stranded DNA binding site of PcrA helicase. *Proceedings of the National Academy of Sciences of the United States of America* **98**, 8381–8387.

513. Soultanas, P., and Wigley, D. B. (2000). DNA helicases: "inching forward". *Current Opinion in Structural Biology* **10**, 124–128.

514. Soultanas, P., and Wigley, D. B. (2001). Unwinding the "Gordian knot" of helicase action. *Trends in Biochemical Sciences* **26**, 47–54.

515. Ivessa, A. S., Zhou, J. Q., Schulz, V. P., Monson, E. K., and Zakian, V. A. (2002). Saccharomyces Rrm3p, a 5′ to 3′ DNA helicase that promotes replication fork progression through telomeric and subtelomeric DNA. *Genes and Development* **16**, 1383–1396.

516. Singleton, M. R., Dillingham, M. S., Gaudier, M., Kowalczykowski, S. C., and Wigley, D. B. (2004). Crystal structure of RecBCD enzyme reveals a machine for processing DNA breaks. *Nature* **432**, 187–193.

517. Barry, J., and Alberts, B. (1994). A role for two DNA helicases in the replication of T4 bacteriophage DNA. *The Journal of Biological Chemistry* **269**, 33063–33068.

518. Gauss, P., Park, K., Spencer, T. E., and Hacker, K. J. (1994). DNA helicase requirements for DNA replication during bacteriophage T4 infection. *Journal of Bacteriology* **176**, 1667–1672.

519. Raney, K. D., and Benkovic, S. J. (1995). Bacteriophage T4 Dda helicase translocates in a unidirectional fashion on single-stranded DNA. *The Journal of Biological Chemistry* **270**, 22236–22242.

520. Morris, P. D., Tackett, A. J., and Raney, K. D. (2001). Biotin-streptavidin-labeled oligonucleotides as probes of helicase mechanisms. *Methods* **23**, 149–159.

521. Morris, P. D., and Raney, K. D. (1999). DNA helicases displace streptavidin from biotin-labeled oligonucleotides. *Biochemistry* **38**, 5164–5171.

522. Taylor, A. F., and Smith, G. R. (1995). Monomeric RecBCD enzyme binds and unwinds DNA. *The Journal of Biological Chemistry* **270**, 24451–24458.

523. Nanduri, B., Byrd, A. K., Eoff, R. L., Tackett, A. J., and Raney, K. D. (2002). Pre-steady-state DNA unwinding by bacteriophage T4 Dda helicase reveals a monomeric molecular motor. *Proceedings of the National Academy of Sciences of the United States of America* **99**, 14722–14727.

524. Cordin, O., Banroques, J., Tanner, N. K., and Linder, P. (2006). The DEAD-box protein family of RNA helicases. *Gene* **367**, 17–37.

525. Bennett, R. J., and Keck, J. L. (2004). Structure and function of RecQ DNA helicases. *Critical Reviews in Biochemistry and Molecular Biology* **39**, 79–97.

526. Flaus, A., and Owen-Hughes, T. (2004). Mechanisms for ATP-dependent chromatin remodelling: farewell to the tuna-can octamer? *Current Opinion in Genetics and Development* **14**, 165–173.

527. Flaus, A., Martin, D. M., Barton, G. J., and Owen-Hughes, T. (2006). Identification of multiple distinct Snf2 subfamilies with conserved structural motifs. *Nucleic Acids Research* **34**, 2887–2905.

528. Linder, P., and Lasko, P. (2006). Bent out of shape: RNA unwinding by the DEAD-box helicase Vasa. *Cell* **125**, 219–221.

529. Sung, P., Bailly, V., Weber, C., Thompson, L. H., Prakash, L., and Prakash, S. (1993). Human xeroderma pigmentosum group D gene encodes a DNA helicase. *Nature* **365**, 852–855.

530. Pradhan, A., and Tuteja, R. (2007). Bipolar, dual Plasmodium falciparum helicase 45 expressed in the intraerythrocytic developmental cycle is required for parasite growth. *Journal of Molecular Biology* **373**, 268–281.

531. Pang, P. S., Jankowsky, E., Planet, P. J., and Pyle, A. M. (2002). The hepatitis C viral NS3 protein is a processive DNA helicase with cofactor enhanced RNA unwinding. *The EMBO Journal* **21**, 1168–1176.

532. Tai, C. L., Chi, W. K., Chen, D. S., and Hwang, L. H. (1996). The helicase activity associated with hepatitis C virus nonstructural protein 3 (NS3). *Journal of Virology* **70**, 8477–8484.

533. Appleby, T. C., Anderson, R., Fedorova, O., Pyle, A. M., Wang, R., Liu, X., Brendza, K. M., and Somoza, J. R. (2011). Visualizing ATP-Dependent RNA Translocation by the NS3 Helicase from HCV. *Journal of Molecular Biology* **405**, 1139–1153.

534. Dumont, S., Cheng, W., Serebrov, V., Beran, R. K., Tinoco, I., Jr., Pyle, A. M., and Bustamante, C. (2006). RNA translocation and unwinding mechanism of HCV NS3 helicase and its coordination by ATP. *Nature* **439**, 105–108.

535. Tackett, A. J., Chen, Y., Cameron, C. E., and Raney, K. D. (2005). Multiple full-length NS3 molecules are required for optimal unwinding of

oligonucleotide DNA *in vitro*. *The Journal of Biological Chemistry* **280**, 10797–10806.

536. Kim, J. L., Morgenstern, K. A., Griffith, J. P., Dwyer, M. D., Thomson, J. A., Murcko, M. A., Lin, C., and Caron, P. R. (1998). Hepatitis C virus NS3 RNA helicase domain with a bound oligonucleotide: The crystal structure provides insights into the mode of unwinding. *Structure* **6**, 89–100.
537. Mackintosh, S. G., Lu, J. Z., Jordan, J. B., Harrison, M. K., Sikora, B., Sharma, S. D., Cameron, C. E., Raney, K. D., and Sakon, J. (2006). Structural and biological identification of residues on the surface of NS3 helicase required for optimal replication of the hepatitis C virus. *The Journal of Biological Chemistry* **281**, 3528–3535.
538. Kim, J. W., Seo, M. Y., Shelat, A., Kim, C. S., Kwon, T. W., Lu, H. H., Moustakas, D. T., Sun, J., and Han, J. H. (2003). Structurally conserved amino Acid w501 is required for RNA helicase activity but is not essential for DNA helicase activity of hepatitis C virus NS3 protein. *Journal of Virology* **77**, 571–582.
539. Tai, C. L., Pan, W. C., Liaw, S. H., Yang, U. C., Hwang, L. H., and Chen, D. S. (2001). Structure-based mutational analysis of the hepatitis C virus NS3 helicase. *Journal of Virology* **75**, 8289–8297.
540. Paolini, C., Lahm, A., De Francesco, R., and Gallinari, P. (2000). Mutational analysis of hepatitis C virus NS3-associated helicase. *Journal of General Virology* **81**, 1649–1658.
541. Sengoku, T., Nureki, O., Nakamura, A., Kobayashi, S., and Yokoyama, S. (2006). Structural basis for RNA unwinding by the DEAD-box protein Drosophila Vasa. *Cell* **125**, 287–300.
542. Serebrov, V., and Pyle, A. M. (2004). Periodic cycles of RNA unwinding and pausing by hepatitis C virus NS3 helicase. *Nature* **430**, 476–480.
543. Jankowsky, E., Gross, C. H., Shuman, S., and Pyle, A. M. (2000). The DExH protein NPH-II is a processive and directional motor for unwinding RNA. *Nature* **403**, 447–451.
544. Kawaoka, J., Jankowsky, E., and Pyle, A. M. (2004). Backbone tracking by the SF2 helicase NPH-II. *Nature Structural and Molecular Biology* **11**, 526–530.
545. Beran, R. K., Bruno, M. M., Bowers, H. A., Jankowsky, E., and Pyle, A. M. (2006). Robust translocation along a molecular monorail: The NS3 helicase from hepatitis C virus traverses unusually large disruptions in its track. *Journal of Molecular Biology* **358**, 974–982.
546. Levin, M. K., Gurjar, M. M., and Patel, S. S. (2003). ATP binding modulates the nucleic acid affinity of hepatitis C virus helicase. *The Journal of Biological Chemistry* **278**, 23311–23316.

547. Levin, M. K., Gurjar, M., and Patel, S. S. (2005). A Brownian motor mechanism of translocation and strand separation by hepatitis C virus helicase. *Nature Structural and Molecular Biology* **12**, 429–435.

548. Firman, K., and Szczelkun, M. D. (2000). Measuring motion on DNA by the type I restriction endonuclease EcoR124I using triplex displacement. *The EMBO Journal* **19**, 2094–2102.

549. Thoma, N. H., Czyzewski, B. K., Alexeev, A. A., Mazin, A. V., Kowalczykowski, S. C., and Pavletich, N. P. (2005). Structure of the SWI2/SNF2 chromatin-remodeling domain of eukaryotic Rad54. *Nature Structural and Molecular Biology* **12**, 350–356.

550. Durr, H., Korner, C., Muller, M., Hickmann, V., and Hopfner, K. P. (2005). X-ray structures of the Sulfolobus solfataricus SWI2/SNF2 ATPase core and its complex with DNA. *Cell* **121**, 363–373.

551. Singleton, M. R., Scaife, S., and Wigley, D. B. (2001). Structural analysis of DNA replication fork reversal by RecG. *Cell* **107**, 79–89.

552. Flaus, A., and Owen-Hughes, T. (2001). Mechanisms for ATP-dependent chromatin remodelling. *Current Opinion in Genetics and Development* **11**, 148–154.

553. Brower-Toland, B. D., Smith, C. L., Yeh, R. C., Lis, J. T., Peterson, C. L., and Wang, M. D. (2002). Mechanical disruption of individual nucleosomes reveals a reversible multistage release of DNA. *Proceedings of the National Academy of Sciences of the United States of America* **99**, 1960–1965.

554. Langst, G., Bonte, E. J., Corona, D. F., and Becker, P. B. (1999). Nucleosome movement by CHRAC and ISWI without disruption or trans-displacement of the histone octamer. *Cell* **97**, 843–852.

555. Saha, A., Wittmeyer, J., and Cairns, B. R. (2005). Chromatin remodeling through directional DNA translocation from an internal nucleosomal site. *Nature Structural and Molecular Biology* **12**, 747–755.

556. Chaban, Y., Ezeokonkwo, C., Chung, W. H., Zhang, F., Kornberg, R. D., Maier-Davis, B., Lorch, Y., and Asturias, F. J. (2008). Structure of a RSC-nucleosome complex and insights into chromatin remodeling. *Nature Structural and Molecular Biology* **15**, 1272–1277.

557. Shukla, M. S., Syed, S. H., Montel, F., Faivre-Moskalenko, C., Bednar, J., Travers, A., Angelov, D., and Dimitrov, S. (2010). Remosomes: RSC generated non-mobilized particles with approximately 180 bp DNA loosely associated with the histone octamer. *Proceedings of the National Academy of Sciences of the United States of America* **107**, 1936–1941.

558. Clapier, C. R., and Cairns, B. R. (2009). The biology of chromatin remodeling complexes. *Annual Review of Biochemistry* **78**, 273–304.

559. Sirinakis, G., Clapier, C. R., Gao, Y., Viswanathan, R., Cairns, B. R., and Zhang, Y. (2011). The RSC chromatin remodelling ATPase translocates DNA with high force and small step size. *The EMBO Journal* **30**, 2364–2372.

560. van Noort, J., van der Heijden, T., Dutta, C. F., Firman, K., and Dekker, C. (2004). Initiation of translocation by Type I restriction-modification enzymes is associated with a short DNA extrusion. *Nucleic Acids Research* **32**, 6540–6547.

561. Davey, M. J., and O'Donnell, M. (2003). Replicative helicase loaders: Ring breakers and ring makers. *Current Biology* **13**, R594–R596.

562. Gorbalenya, A. E., Koonin, E. V., and Wolf, Y. I. (1990). A new superfamily of putative NTP-binding domains encoded by genomes of small DNA and RNA viruses. *FEBS Letters* **262**, 145–148.

563. Yoshimoto, K., Arora, K., and Brooks, C. L. (2010). Hexameric helicase deconstructed: Interplay of conformational changes and substrate coupling. *Biophysical Journal* **98**, 1449–1457.

564. Hickman, A. B., and Dyda, F. (2005). Binding and unwinding: SF3 viral helicases. *Current Opinion in Structural Biology* **15**, 77–85.

565. Enemark, E. J., and Joshua-Tor, L. (2006). Mechanism of DNA translocation in a replicative hexameric helicase. *Nature* **442**, 270–275.

566. Ilyina, T. V., Gorbalenya, A. E., and Koonin, E. V. (1992). Organization and evolution of bacterial and bacteriophage primase-helicase systems. *Journal of Molecular Evolution* **34**, 351–357.

567. Toth, E. A., Li, Y., Sawaya, M. R., Cheng, Y., and Ellenberger, T. (2003). The crystal structure of the bifunctional primase-helicase of bacteriophage T7. *Molecular Cell* **12**, 1113–1123.

568. Sawaya, M. R., Guo, S., Tabor, S., Richardson, C. C., and Ellenberger, T. (1999). Crystal structure of the helicase domain from the replicative helicase-primase of bacteriophage T7. *Cell* **99**, 167–177.

569. Singleton, M. R., Sawaya, M. R., Ellenberger, T., and Wigley, D. B. (2000). Crystal structure of T7 gene 4 ring helicase indicates a mechanism for sequential hydrolysis of nucleotides. *Cell* **101**, 589–600.

570. Washington, M. T., Rosenberg, A. H., Griffin, K., Studier, F. W., and Patel, S. S. (1996). Biochemical analysis of mutant T7 primase/helicase proteins defective in DNA binding, nucleotide hydrolysis, and the coupling of hydrolysis with DNA unwinding. *The Journal of Biological Chemistry* **271**, 26825–26834.

571. Gogol, E. P., Seifried, S. E., and von Hippel, P. H. (1991). Structure and assembly of the *Escherichia coli* transcription termination factor rho

and its interaction with RNA. I. Cryoelectron microscopic studies. *Journal of Molecular Biology* **221**, 1127–1138.

572. Skordalakes, E., and Berger, J. M. (2003). Structure of the Rho transcription terminator: Mechanism of mRNA recognition and helicase loading. *Cell* **114**, 135–146.

573. Thomsen, N. D., and Berger, J. M. (2009). Running in reverse: The structural basis for translocation polarity in hexameric helicases. *Cell* **139**, 523–534.

574. Labib, K., Tercero, J. A., and Diffley, J. F. (2000). Uninterrupted MCM2-7 function required for DNA replication fork progression. *Science* **288**, 1643–1647.

575. Chong, J. P., Hayashi, M. K., Simon, M. N., Xu, R. M., and Stillman, B. (2000). A double-hexamer archaeal minichromosome maintenance protein is an ATP-dependent DNA helicase. *Proceedings of the National Academy of Sciences of the United States of America* **97**, 1530–1535.

576. Kaplan, D. L., and O'Donnell, M. (2004). Twin DNA pumps of a hexameric helicase provide power to simultaneously melt two duplexes. *Molecular Cell* **15**, 453–465.

577. Fletcher, R. J., Bishop, B. E., Leon, R. P., Sclafani, R. A., Ogata, C. M., and Chen, X. S. (2003). The structure and function of MCM from archaeal M. Thermoautotrophicum. *Nature Structural Biology* **10**, 160–167.

578. West, S. C. (1996). The RuvABC proteins and Holliday junction processing in *Escherichia coli*. *Journal of Bacteriology* **178**, 1237–1241.

579. Matias, P. M., Gorynia, S., Donner, P., and Carrondo, M. A. (2006). Crystal structure of the human AAA$^+$ protein RuvBL1. *The Journal of Biological Chemistry* **281**, 38918–38929.

580. Lesterlin, C., Barre, F. X., and Cornet, F. (2004). Genetic recombination and the cell cycle: What we have learned from chromosome dimers. *Molecular Microbiology* **54**, 1151–1160.

581. Begg, K. J., Dewar, S. J., and Donachie, W. D. (1995). A new *Escherichia coli* cell division gene, ftsK. *Journal of Bacteriology* **177**, 6211–6222.

582. Draper, G. C., McLennan, N., Begg, K., Masters, M., and Donachie, W. D. (1998). Only the N-terminal domain of FtsK functions in cell division. *Journal of Bacteriology* **180**, 4621–4627.

583. Yates, J., Aroyo, M., Sherratt, D. J., and Barre, F. X. (2003). Species specificity in the activation of Xer recombination at dif by FtsK. *Molecular Microbiology* **49**, 241–249.

584. Dubarry, N., Possoz, C., and Barre, F.-X. (2010). Multiple regions along the *Escherichia coli* FtsK protein are implicated in cell division. *Molecular Microbiology* **78**, 1088–1100.

585. Aussel, L., Barre, F. X., Aroyo, M., Stasiak, A., Stasiak, A. Z., and Sherratt, D. (2002). FtsK Is a DNA motor protein that activates chromosome dimer resolution by switching the catalytic state of the XerC and XerD recombinases. *Cell* **108**, 195–205.

586. Massey, T. H., Mercogliano, C. P., Yates, J., Sherratt, D. J., and Lowe, J. (2006). Double-stranded DNA translocation: Structure and mechanism of hexameric FtsK. *Molecular Cell* **23**, 457–469.

587. Graham, J. E., Sherratt, D. J., and Szczelkun, M. D. (2010). Sequence-specific assembly of FtsK hexamers establishes directional translocation on DNA. *Proceedings of the National Academy of Sciences of the United States of America* **107**, 20263–20268.

588. Moncalian, G., Cabezon, E., Alkorta, I., Valle, M., Moro, F., Valpuesta, J. M., Goni, F. M., and de La Cruz, F. (1999). Characterization of ATP and DNA binding activities of TrwB, the coupling protein essential in plasmid R388 conjugation. *The Journal of Biological Chemistry* **274**, 36117–36124.

589. Gomis-Ruth, F. X., and Coll, M. (2001). Structure of TrwB, a gatekeeper in bacterial conjugation. *The International Journal of Biochemistry and Cell Biology* **33**, 839–843.

590. Mancini, E. J., Kainov, D. E., Grimes, J. M., Tuma, R., Bamford, D. H., and Stuart, D. I. (2004). Atomic snapshots of an RNA packaging motor reveal conformational changes linking ATP hydrolysis to RNA translocation. *Cell* **118**, 743–755.

591. Abrahams, J. P., Leslie, A. G. W., Lutter, R., and Walker, J. E. (1994). Structure at 2.8-Angstrom Resolution of F_1-ATPase from Bovine Heart-Mitochondria. *Nature* **370**, 621–628.

592. Hingorani, M. M., Washington, M. T., Moore, K. C., and Patel, S. S. (1997). The dTTPase mechanism of T7 DNA helicase resembles the binding change mechanism of the F_1-ATPase. *Proceedings of the National Academy of Sciences of the United States of America* **94**, 5012–5017.

593. Liao, J. C., Jeong, Y. J., Kim, D. E., Patel, S. S., and Oster, G. (2005). Mechanochemistry of T7 DNA helicase. *Journal of Molecular Biology* **350**, 452–475.

594. Adelman, J. L., Jeong, Y. J., Liao, J. C., Patel, G., Kim, D. E., Oster, G., and Patel, S. S. (2006). Mechanochemistry of transcription termination factor Rho. *Molecular Cell* **22**, 611–621.

595. Crampton, D. J., Mukherjee, S., and Richardson, C. C. (2006). DNA-induced switch from independent to sequential dTTP hydrolysis in the bacteriophage T7 DNA helicase. *Molecular Cell* **21**, 165–174.

596. Martin, A., Baker, T. A., and Sauer, R. T. (2005). Rebuilt AAA$^+$ motors reveal operating principles for ATP-fuelled machines. *Nature* **437**, 1115–1120.

597. Gai, D., Zhao, R., Li, D., Finkielstein, C. V., and Chen, X. S. (2004). Mechanisms of conformational change for a replicative hexameric helicase of SV40 large tumor antigen. *Cell* **119**, 47–60.

598. Enemark, E. J., and Joshua-Tor, L. (2008). On helicases and other motor proteins. *Current Opinion in Structural Biology* **18**, 243–257.

599. Enemark, E. J., and Joshua-Tor, L. (2008). On helicases and other motor proteins. *Current Opinion in Structural Biology* **18**, 243–257.

600. Dittrich, M., and Schulten, K. (2006). PcrA helicase, a prototype ATP-driven molecular motor. *Structure* **14**, 1345–1353.

601. Lee, J. Y., and Yang, W. (2006). UvrD helicase unwinds DNA one base pair at a time by a two-part power stroke. *Cell* **127**, 1349–1360.

602. Lenzen, C. U., Steinmann, D., Whiteheart, S. W., and Weis, W. I. (1998). Crystal structure of the hexamerization domain of N-ethylmaleimide-sensitive fusion protein. *Cell* **94**, 525–536.

603. Yu, R. C., Hanson, P. I., Jahn, R., and Brunger, A. T. (1998). Structure of the ATP-dependent oligomerization domain of N-ethylmaleimide sensitive factor complexed with ATP. *Nature Structural Biology* **5**, 803–811.

604. Bono, F., Ebert, J., Lorentzen, E., and Conti, E. (2006). The crystal structure of the exon junction complex reveals how it maintains a stable grip on mRNA. *Cell* **126**, 713–725.

605. Lucic, B., Zhang, Y., King, O., Mendoza-Maldonado, R., Berti, M., Niesen, F. H., Burgess-Brown, N. A., Pike, A. C. W., Cooper, C. D. O., Gileadi, O., and Vindigni, A. (2011). A prominent β-hairpin structure in the winged-helix domain of RECQ1 is required for DNA unwinding and oligomer formation. *Nucleic Acids Research* **39**, 1703–1717.

606. Neuwald, A. F., Aravind, L., Spouge, J. L., and Koonin, E. V. (1999). AAA$^+$: A class of chaperone-like ATPases associated with the assembly, operation, and disassembly of protein complexes. *Genome Research* **9**, 27–43.

607. Sekimizu, K., Bramhill, D., and Kornberg, A. (1987). ATP activates dnaA protein in initiating replication of plasmids bearing the origin of the *E. coli* chromosome. *Cell* **50**, 259–265.

608. Erzberger, J. P., Pirruccello, M. M., and Berger, J. M. (2002). The structure of bacterial DnaA: Implications for general mechanisms underlying DNA replication initiation. *The EMBO Journal* **21**, 4763–4773.

609. Yodh, J. G., Schlierf, M., and Ha, T. (2010). Insight into helicase mechanism and function revealed through single-molecule approaches. *Quarterly Reviews of Biophysics* **43**, 185–218.

610. Myong, S., and Ha, T. (2010). Stepwise translocation of nucleic acid motors. *Current Opinion in Structural Biology* **20**, 121–127.

611. Manosas, M., Xi, X. G., Bensimon, D., and Croquette, V. (2010). Active and passive mechanisms of helicases. *Nucleic Acids Research* **38**, 5518–5526.

612. Chen, Y., Guzik, S., Sumner, J. P., Moreland, J., and Koretsky, A. P. (2011). Magnetic manipulation of actin orientation, polymerization, and gliding on myosin using superparamagnetic iron oxide particles. *Nanotechnology* **22**, 065101.

613. Mendoza-Maldonado, R., Faoro, V., Bajpai, S., Berti, M., Odreman, F., Vindigni, M., Ius, T., Ghasemian, A., Bonin, S., Skarp, M., Stanta, G., and Vindigni, A. (2011). The human RECQ1 helicase is highly expressed in glioblastoma and plays an important role in tumor cell proliferation. *Molecular Cancer* **10**, 83.

614. Kerssemakers, J., Ionov, L., Queitsch, U., Luna, S., Hess, H., and Diez, S. (2009). 3D nanometer tracking of motile microtubules on reflective surfaces. *Small* **5**, 1732–1737.

615. Heuston, E., Bronner, C., Kull, F., and Endow, S. (2010). A kinesin motor in a force- producing conformation. *BMC Structural Biology* **10**, 19.

616. Fischer, T., Agarwal, A., and Hess, H. (2009). A smart dust biosensor powered by kinesin motors. *Nature Nanotechnology* **4**, 162–166.

617. Greene, A. C., Trent, A. M., and Bachand, G. D. (2008). Controlling kinesin motor proteins in nanoengineered systems through a metal-binding on/off switch. *Biotechnology and Bioengineering* **101**, 478–486.

618. Hyeon, C., and Onuchic, J. N. (2007). Mechanical control of the directional stepping dynamics of the kinesin motor. *Proceedings of the National Academy of Sciences of the United States of America* **104**, 17382–17387.

619. Buonocore, A., Caputo, L., Ishii, Y., Pirozzi, E., Yanagida, T., and Ricciardi, L. M. (2005). On Myosin II dynamics in the presence of external loads. *Biosystems* **81**, 163–175.

620. Straight, A. F., Field, C. M., and Mitchison, T. J. (2005). Anillin binds nonmuscle myosin II and regulates the contractile ring. *Molecular Biology of the Cell* **16**, 193–201.

621. Lister, I., Roberts, R., Schmitz, S., Walker, M., Trinick, J., Veigel, C., Buss, F., and Kendrick-Jones, J. (2004). Myosin VI: A multifunctional motor. *Biochemical Society Transaction* **32**, 685–688.

622. Kumari, A., Minko, I. G., Smith, R. L., Lloyd, R. S., and McCullough, A. K. (2010). Modulation of UvrD helicase activity by covalent DNA-protein cross-links. *The Journal of Biological Chemistry* **285**, 21313.

623. Boubakri, H., de Septenville, A. L., Viguera, E., and Michel, B. (2010). The helicases DinG, Rep and UvrD cooperate to promote replication across transcription units *in vivo*. *The EMBO Journal* **29**, 145–157.

624. Plank, J., and Hsieh, T. S. (2009). Helicase-appended topoisomerases: New insight into the mechanism of directional strand transfer. *The Journal of Biological Chemistry* **284**, 30737.

625. Seidel, R., Bloom, J. G. P., van Noort, J., Dutta, C. F., Dekker, N. H., Firman, K., Szczelkun, M. D., and Dekker, C. (2005). Dynamics of initiation, termination and reinitiation of DNA translocation by the motor protein EcoR124I. *European Molecular Biology Organisation Journal* **24**, 4188–4197.

626. Hess, H., Bachand, G. D., and Vogel, V. (2004). Powering nanodevices with biomolecular motors. *Chemistry* **10**, 2110–2116.

627. Drexler, K. E. (1992). *Nanosystems: Molecular Machinery, Manufacturing and Computation*, John Wiley and Sons, Inc, USA.

628. Drexler, K. E. (1999). Building molecular machine systems. *Trends in Biotechnology* **17**, 5–7.

629. Pease, A. R., Jeppesen, J. O., Stoddart, J. F., Luo, Y., Collier, C. P., and Heath, J. R. (2001). Switching devices based on interlocked molecules†. *Accounts of Chemical Research* **34**, 433–444.

630. Schalley, C. A., Beizai, K., and Vögtle, F. (2001). On the way to rotaxane-based molecular motors: Studies in molecular mobility and topological chirality†. *Accounts of Chemical Research* **34**, 465–476.

631. Collier, C. P., Mattersteig, G., Wong, E. W., Luo, Y., Beverly, K., Sampaio, J., Raymo, F. M., Stoddart, J. F., and Heath, J. R. (2000). A [2]catenane-based solid state electronically reconfigurable switch. *Science* **289**, 1172–1175.

632. Collier, C. P., Jeppesen, J. O., Luo, Y., Perkins, J., Wong, E. W., Heath, J. R., and Stoddart, J. F. (2001). Molecular-based electronically switchable tunnel junction devices. *Journal of the American Chemical Society* **123**, 12632–12641.

633. Reches, M., and Gazit, E. (2003). Casting metal nanowires within discrete self-assembled peptide nanotubes. *Science* **300**, 625–627.

634. Fahmi, A. W., Braun, H. G., and Stamm, M. (2003). Fabrication of metallized nanowires from self-assembled diblock copolymer templates. *Advanced Materials* **15**, 1201–1204.

635. Tilke, A., Pescini, L., Erbe, A., Lorenz, H., and Blick, R. H. (2002). Electron-phonon interaction in suspended highly doped silicon nanowires. *Nanotechnology* **13**, 491–494.

636. Collin, J.-P., Dietrich-Buchecker, C., Gaviña, P., Jimenez-Molero, M. C., and Sauvage, J.-P. (2001). Shuttles and muscles: Linear molecular machines based on transition metals†. *Accounts of Chemical Research* **34**, 477–487.

637. Giljohann, D. A., and Mirkin, C. A. (2009). Drivers of biodiagnostic development. *Nature* **462**, 461–464.

638. Soong, R. K., Bachand, G. D., Neves, H. P., Olkhovets, A. G., Craighead, H. G., and Montemagno, C. D. (2000). Powering an inorganic nanodevice with a biomolecular motor. *Science* **290**, 1555–1558.

639. Agarwal, A., and Hess, H. (2010). Biomolecular motors at the intersection of nanotechnology and polymer science. *Progress in Polymer Science* **35**, 252–277.

640. Xu, Z., Wen, J.-K., Liu, C., Liu, J.-S., Du, L.-Q., and Wang, L.-D. (2009). Research on forming and application of U-form glass micro-nanofluidic chip with long nanochannels. *Microfluidics and Nanofluidics* **7**, 423–429.

641. Zhang, L., Gu, F., Tong, L., and Yin, X. (2008). Simple and cost-effective fabrication of two-dimensional plastic nanochannels from silica nanowire templates. *Microfluidics and Nanofluidics* **5**, 727–732.

642. Sparreboom, W., van den Berg, A., and Eijkel, J. C. (2009). Principles and applications of nanofluidic transport. *Nature Nanotechnology* **4**, 713–720.

643. Malcos, J., and Hancock, W. (2011). Engineering tubulin: Microtubule functionalization approaches for nanoscale device applications. *Applied Microbiology and Biotechnology* **90**, 1–10.

644. Clemmens, J., Hess, H., Lipscomb, R., Hanein, Y., Bohringer, K. F., Matzke, C. M., Bachand, G. D., Bunker, B. C., and Vogel, V. (2003). Mechanisms of microtubule guiding on microfabricated kinesin-coated surfaces: Chemical and topographic surface patterns. *Langmuir* **19**, 10967–10974.

645. Hirabayashi, M., Taira, S., Kobayashi, S., Konishi, K., Katoh, K., Hiratsuka, Y., Kodaka, M., Uyeda, T. Q., Yumoto, N., and Kubo, T. (2006). Malachite green-conjugated microtubules as mobile bioprobes selective for malachite green aptamers with capturing/releasing ability. *Biotechnology and Bioengineering* **94**, 473–480.

646. Ramachandran, S., Ernst, K. H., Bachand, G. D., Vogel, V., and Hess, H. (2006). Selective loading of kinesin-powered molecular shuttles with protein cargo and its application to biosensing. *Small* **2**, 330–334.

647. Brunner, C., Wahnes, C., and Vogel, V. (2007). Cargo pick-up from engineered loading stations by kinesin driven molecular shuttles. *Lab on a Chip* **7**, 1263–1271.

648. Hess, H., Clemmens, J., Qin, D., Howard, J., and Vogel, V. (2001). Light-controlled molecular shuttles made from motor proteins carrying cargo on engineered surfaces. *Nano Letters* **1**, 235–239.

649. Suzuki, N., Miyata, H., Ishiwata, S., and Kinosita, K. (1996). Preparation of bead-tailed actin filaments: Estimation of the torque produced by the sliding force in an *in vitro* motility assay. *Biophysical Journal* **70**, 401–408.

650. Tucker, R., Katira, P., and Hess, H. (2007). Herding nanotransporters: Localized activation via release and sequestration of control molecules. *Nano Letters* **8**, 221–226.

651. Ionov, L., Stamm, M., and Diez, S. (2006). Reversible switching of microtubule motility using thermoresponsive polymer surfaces. *Nano Letters* **6**, 1982–1987.

652. Korten, T., Månsson, A., and Diez, S. (2010). Towards the application of cytoskeletal motor proteins in molecular detection and diagnostic devices. *Current Opinion in Biotechnology* **21**, 477–488.

653. Stanley, L. K., and Szczelkun, M. D. (2006). Direct and random routing of a molecular motor protein at a DNA junction. *Nucleic Acids Research* **34**, 4387–4394.

654. Sanderson, K. (2010). Bioengineering: What to make with DNA origami. *Nature* **464**, 158–159.

655. Firman, K., and Szczelkun, M. (2000). Measuring motion on DNA by the Type I restriction endonuclease EcoR124I using triplex dissociation. *European Molecular Biology Organisation Journal* **19**, 2094–2102.

656. Chaves, R. C., Bensimon, D., and Freitas, P. P. (2011). Single molecule actuation and detection on a lab-on-a-chip magnetoresistive platform. *Journal of Applied Physics* **109**.

657. Koster, D. A., Palle, K., Bot, E. S. M., Bjornsti, M.-A., and Dekker, N. H. (2007). Antitumour drugs impede DNA uncoiling by topoisomerase I. *Nature* **448**, 213–217.

658. Kwa, F. A. A., Balcerczyk, A., Licciardi, P., El-Osta, A., and Karagiannis, T. C. (2011). Chromatin modifying agents: The cutting edge of anticancer therapy. *Drug Discovery Today* **16**, 543–547.

659. Dekker, F. J., and Haisma, H. J. (2009). Histone acetyl transferases as emerging drug targets. *Drug Discovery Today* **14**, 942–948.

660. Ma, J., Xu, Y., Zhang, Q., Zha, L., and Liang, B. (2007). Preparation and characterization of pH- and temperature-responsive semi-IPN hydrogels of carboxymethyl chitosan with poly (N-isopropyl acrylamide) crosslinked by clay. *Colloid and Polymer Science* **285**, 479–484.

661. Mohan, Y. M., and Geckeler, K. E. (2007). Polyampholytic hydrogels: Poly(N-isopropylacrylamide)-based stimuli-responsive networks with poly(ethyleneimine). *Reactive and Functional Polymers* **67**, 144–155.

662. Morishima, Y. (2007). Thermally responsive polymer vesicles. *Angewandte Chemie International Edition in English* **46**, 1370–1372.

663. Filipcsei, G., Csetneki, I., Szilagyi, A., and Zrinyi, M. (2007). Magnetic field-responsive smart polymer composites. *Advances in Polymer Science* **206**, 137–190.

664. Xu, Y., Shi, L., Ma, R., Zhang, W., An, Y., and Zhu, X. X. (2007). Synthesis and micellization of thermo- and pH-responsive block copolymer of poly(N-isopropylacrylamide)-block-poly(4-vinylpyridine). *Polymer* **48**, 1711–1717.

665. Guo, B., Yuan, J., Yao, L., and Gao, Q. (2007). Preparation and release profiles of pH/temperature-responsive carboxymethyl chitosan/P(2-(dimethylamino) ethyl methacrylate) semi-IPN amphoteric hydrogel. *Colloid and Polymer Science* **285**, 665–671.

666. Pennadam, S., Lavigne, M. D., Dutta, C. F., Firman, K., Mernagh, D., Górecki, D. C., and Cameron, A. (2004). Control of a multi-subunit DNA motor by a thermo responsive polymer switch. *Journal of the American Chemical Society* **126**, 13208–13209.

667. Slocik, J. M., Tam, F., Halas, N. J., and Naik, R. R. (2007). Peptide-assembled optically responsive nanoparticle complexes. *Nano Letters* **7(4)**, 1054–1058.

668. Pistol, C., and Dwyer, C. (2007). Scalable, low-cost, hierarchical assembly of programmable DNA nanostructures. *Nanotechnology* **18**, 125305.

669. Liedl, T., Sobey, T. L., and Simmel, F. C. (2007). DNA-based nanodevices. *Nano Today* **2**, 36–41.

670. Regis, E. (1995). *Nano: The Emerging Science of Nanotechnology*, 1st ed, Little, Brown and Company.

671. Drexler, K. E., Peterson, C., and Pergamit, G. (1991). *Unbounding the Future: The Nanotechnology Revolution*, Diane Pub Co, Morrow.

672. Schwartz, P. V. (2002). Molecular transport from an atomic force microscope tip: A comparative study of dip-pen nanolithography. *Langmuir* **18**, 4041–4046.

673. Liu, G.-Y., and Amro, N. A. (2002). Positioning protein molecules on surfaces: A nanoengineering approach to supramolecular chemistry. *Proceedings of the National Academy of Sciences of the United States of America* **99**, 5165–5170.

674. Conibear, P. B., Kuhlman, P. A., and Bagshaw, C. R. (1998). Measurement of ATPase activities of myosin at the level of tracks and single molecules. *Advances in Experimental Medicine and Biology* **453**, 15–26; discussion 26–27.

675. Overly, C. C., Lee, K. D., Berthiaume, E., and Hollenbeck, P. J. (1995). Quantitative measurement of intraorganelle pH in the endosomal-lysosomal pathway in neurons by using ratiometric imaging with pyranine. *Proceedings of the National Academy of Sciences of the United States of America* **92**, 3156–3160.

676. Mullard, A. (2011). Could pharma open its drug freezers? *Nature Reviews Drug Discovery* **10**, 399–400.

Index